Extraterrestrial Remote Sensing and Climate Change

Extraterrestrial Remote Sensing and Climate Change

Saumitra Mukherjee
Jawaharlal Nehru University
New Delhi
India

WILEY

This edition first published 2023

Registered Office
John Wiley & Sons, Inc., 111 River Street, Hoboken, NJ 07030, USA

Editorial Office
Boschstr. 12, 69469 Weinheim, Germany

For details of our global editorial offices, customer services, and more information about Wiley products visit us at www.wiley.com.

Wiley also publishes its books in a variety of electronic formats and by print-on-demand. Some content that appears in standard print versions of this book may not be available in other formats.

Library of Congress Cataloging-in-Publication Data

Names: Mukherjee, Saumitra, author. | John Wiley & Sons, publisher.
Title: Extraterrestrial remote sensing and climate change / Saumitra Mukherjee.
Description: Hoboken, NJ : Wiley, 2023. | Includes bibliographical references and index.
Identifiers: LCCN 2022026423 (print) | LCCN 2022026424 (ebook) | ISBN 9781119164623 (cloth) | ISBN 9781119164630 (adobe pdf) | ISBN 9781119164647 (epub)
Subjects: LCSH: Climatic changes–Remote sensing. | Climatic changes–Effect of solar activity on. | Environmental monitoring–Remote sensing. | Solar system–Remote sensing. | Artificial satellites in remote sensing.
Classification: LCC QC903 .M85 2023 (print) | LCC QC903 (ebook) | DDC 551.60285–dc23/eng20221026
LC record available at https://lccn.loc.gov/2022026423
LC ebook record available at https://lccn.loc.gov/2022026424

Cover design: Wiley
Cover image: © muratart/Shutterstock

Set in 9.5/12.5pt STIXTwoText by Straive, Chennai, India

Contents

Preface

We have moved to a time to understand the influence of extraterrestrial changes on the Earth and planetary system of the Sun. Climate change manifestations are manifold and include not only the erratic changes in the atmospheric and oceanic temperature of the Earth but also in the whole environment of the Earth. Important extraterrestrial and local scale changes can be correlated by other planets and their moons by the extremely advanced sensors available in satellites. Extraterrestrial remote sensing is an evolving science meant to infer and address the unsolved problems of the terrestrial system, including mutation of viruses before pandemic outbreaks. Using sensors it is possible to correlate the tectonic activities of the moon and the Earth that hold the key to earthquake prediction possibility in future. In this book we attempt to correlate the magnetic field variations in different planetary systems and helioseismological parameters. Besides the correlation of electron flux and proton flux before earthquakes, global warming, erratic snowfall and rainfall, the cosmic rays from the Sun, and galactic sources have shown different correlation. Space Environment Viewing and Analysis Network (SEVAN) instruments installed at Jawaharlal Nehru University, funded by NASA (United States), show a rise in galactic cosmic rays before sudden rainfall and cold waves, while the cosmic rays from the Sun rise before global warming. Similar correlation has also been observed during the thinning of the Saturn ring after directed solar flares and coronal mass ejection. The future of extraterrestrial remote sensing has the potential to predict changes in the Earth and planetary system for the sustenance of the Earth.

The book *Extraterrestrial Remote Sensing and Climate Change* has been written in 14 chapters with sensor interpretation of satellites of the Sun, the Earth, and other planetary systems to understand natural resources and natural disasters in an interdisciplinary format.

New Delhi
India

Saumitra Mukherjee
Jawaharlal Nehru University

Acknowledgments

The concept of extraterrestrial influence on climate change was developed in my mind more than 50 years ago when I was in my teenage years while discussing with my mother the late Mrs. Bela Rani Mukherjee the change in the universe and the Earth. I conceived the idea about the universe and climate change based on the discussion about the mystique stories told to me by my mother about Swami Vishudhdhanand of Varanasi. Ever-changing Earth should be a part of the changes in the universe; this concept is the key to sustainable development. If the changes are fast and episodic, then the environment is also affected, which may affect the living and nonliving components of the environment. I am thankful to my colleagues and students of the School of Environmental Sciences, Jawaharlal Nehru University, for lending me their moral support during completion of this work. I specially acknowledge Dr. Kamleash Lulla Chief Scientist NASA, Dr. Laszlo Kortvallessey of the Hungarian Academy of Sciences, and Dr. Milan Radovanović of the Serbian Academy of Sciences for fruitful discussion. I have worked in association with these scientists in the last decade and published with them on various aspects of the Sun–Earth–Cosmic connection. I acknowledge NASA, USA, for the use of SOHO satellite data, OMI-KNMI data for the status of air quality concentration, and different cosmic ray data sources available on the website. Special acknowledgments are due to NASA USA for downloading public domain images of different satellites and their data. Financial assistance received from ISRO, India, is also acknowledged. I am thankful to Dr. Barbara J. Thompson and Dr. Nat Gopalaswamy of NASA, USA, for their support. The author acknowledges the grant and support provided by NASA (Influence of Sun and other cosmic factors on Earth's Space Weather Environment, No FA48690714096). I am indebted to Professor Hans J. Hubold, an expert of Peaceful Research of Outer Space of United Nations and Professor Ashot Chillingarian of Cosmic Ray Division, Armenia, for their support. I acknowledge the moral support of my wife Dr. Anita Mukherjee and my children Abhijit and Anisha, who cooperated with me and deserve my appreciation.

June 2022 *Saumitra Mukherjee*

1

Extraterrestrial Remote Sensing

1.1 Introduction

Extraterrestrial remote sensing is a fast developing and emerging science to understand not only climate change but the origin and survival of life in only one place of the universe, that is, the Earth. In the solar system, there is a planetary system but life exists only on the Earth. If we consider the origin and development of the magnetic field in and around the Earth during prebiotic evolution, life came to exist on the Earth in a specific magnetic field condition between Sun and the Earth. The evolution of different natural cycles inside and outside the Earth was governed by the intricacies of Sun–Earth–Cosmic connection. In this consequence, a different climatic niche was evolved. In this universe, everything is changing continuously, so the climate was affected by not only the human activity but it had a larger impact by changes in the extraterrestrial variables, which continuously interact with the Earth system. Climate change is a complex phenomenon and it is required to monitor the changes of anthropogenic and geogenic origin as well as influence from outer space in the ever-changing system. We are aware of the changes in seasonal temperature, wind pattern, and rainfall/snowfall variation in space and time. However, the changes are not uniform globally. Decadal survey on various themes of terrestrial and extraterrestrial origin have been studied; it shows that most of the parameters are not uniform in its character and origin. We are convinced about the effect of climate change and the role of anthropogenic activities in this process but have yet to explain the mechanism of sudden catastrophic changes in the environment. Some changes are having manifestations on different environmental components, but some other changes in atmosphere, biosphere, and geosphere are unexplained. However, a periodic repletion of some of the changes is being studied, which can be correlated with the changes in the Sun, other planets, and stars. A scintillating study was carried out on extraterrestrial remote sensing about sudden snowfall on 24 December 2004 in higher latitude and longitude.

If we jot down the global phenomena, most of the times we either try to see the crisis of the environment of the Earth due to our developmental activities or some calamities, which we cannot predict. The problem is not with the global change; the problem is that we are yet to develop an early warning system. An attempt has been initiated to understand the influence of extraterrestrial changes in the environment of the Earth.

It appears that the Earth and its climate are dependent on the changes in extraterrestrial influence. These extraterrestrial forces are governed by the Sun and Star constellation. Attempts are being made to interpret climate change on the Earth using the terrestrial data of different environmental components; however, the efforts are being made to understand the sudden global changes. The shift of the poles of the Earth to the glacial melting in North and South poles and the high altitude terrain including the Himalayas appear to be under the influence of Sun–Earth–Cosmic

Extraterrestrial Remote Sensing and Climate Change, First Edition. Saumitra Mukherjee.

field changes. The polar position of the Earth has recently shifted. The position of North has shifted toward East, and South poles of the Earth toward West direction. It has been identified by a research team of the University of Leeds, UK, that the North Magnetic Pole of the Earth has shifted away from Canada toward Siberia.

It is postulated that the decade-long low magnetic field between Sun and Earth due to low activity of the Sun is one of the reasons for the pole shift. Further, the swarm of microtremors and a few large earthquakes in Antarctica in 2020–2021 have compelled us to think there is a relationship between tectonic activities and extraterrestrial change. The shift of the Earth's magnetic pole toward Siberia is a phenomenon of change in the earth's interior as well triggered by the changes beyond the Earth. The manifestation of polar wandering is climate change due to the solar and galactic changes.

Polar wandering, terrestrial climate change, and the COVID-19 pandemic outbreak are linked with each other. It is a series of events initiated in outer space in the form of changes in the galactic plane. Impact of the galactic plane has changed the Sun–Earth–Cosmic equilibrium. More galactic cosmic rays and fewer ultraviolet rays from the Sun due to improved ozone layer have changed the mutation probability of viruses. The persistent solar minimum of 2019–2020 is a near replica of the 2009–2010 solar minimum. It is imperative that extraterrestrial change may affect terrestrial environment and health of the living beings of the Earth by evolution of new species of virus. The mechanisms of virus mutations were not very clear till now. Sudden outbreak of pandemics like swine flu of 2009 and COVID-19 have compelled us to see it is possible to develop a correlation of solar-minimum-induced climate change and its implications on initiation or triggering of different types of the viral mutations.

Reduction in greenhouse gases by anthropogenic activities coupled with the low electron flux from the Sun has been responsible for reduction in ultraviolet rays from the Sun. A correlation is being attempted to be established between triggering of COVID-19 pandemics of 2019–2021with an unprecedented solar minimum. During this solar minimum continuous low electron flux, low proton flux and low planetary indices have been observed from the Sun. Sudden changes in electron flux, proton flux, and planetary indices (Kp values) coupled with the high galactic cosmic rays has been observed. Further, the polar wandering has some bearing on the intensity of galactic cosmic ray change.

Extraterrestrial influence on the quality of air has been observed during and after the solar eclipse. Aerosol, carbon dioxide, ozone, nitrous oxide gases, and cloud covers have shown direct relation with the electron flux, proton flux, and planetary indices level fluctuation.

Solar maximum and solar minimum both have specific impact on the environment of the Earth as well as the man-made environment. In 2013, the sudden rise in proton flux had an impact on glacial melting and its release of trapped black carbon in the atmosphere to initiate a local catastrophe in the Himalayan terrain of India. The latest impact of the extraterrestrial change has been observed on the formation of the Southern Ocean in Antarctica in 2021 due to melting and breaking of a huge iceberg A 76. There were two incidents recorded prior to the breaking of the icebergs:

1. Sudden changes in heliophysical parameters: It was recorded that Earth entered a stream of solar wind flowing faster than 500 km/s from an equatorial hole in the Sun's atmosphere. On 20 May there was sparked a minor G1-class geomagnetic storm, which was followed by more aggressive geomagnetic storms.
2. Microtremors in Antarctica: during May 2021, Antarctica It was recorded that was shaken by 6 quakes of magnitude 4.0 or above, 36 quakes between 3.0 and 4.0, and 33 quakes between 2.0 and 3.0.

These two incidents give an indication that the space weather is related with the environment of the Earth.

Subdued solar prominences are manifested as very few or no sunspots and sometimes the Sun is spotless shows that no Sunspots during a specific year on the Earth facing side of the Sun.

Contrary to that, the graph below shows the number of days with a geomagnetic storm per year and how strong those storms were. This gives an idea in which years there were a lot of geomagnetic storms.

For the hundreds of millions of people living in coastal regions around the world, rising seas driven by climate change pose a direct threat. It is imperative to infer that there is a need to plan for the coastal areas of the whole Earth. These measurements of sea-level changes are difficult to get from satellites. In view of this problem, the European Space Agency (ESA) initiated research that demonstrates how processing satellite altimetry data in a specific way now makes it possible to determine sea-level change in coastal areas with millimeter per year accuracy, even if the sea is covered by ice.

As an experiment, the Baltic Sea data were processed by ESA, which shows regional differences in sea-level rise, drastically improving and extending previous sea-level trend maps. It shows that between 1995 and 2019, the sea level has risen at an annual rate of 2–3 mm in the south, along the German and Danish coasts, compared to a few millimeters in the northeast, in the Bay of Bothnia.

Now it is clear that both the ocean and the atmosphere warm together because of climate change. Sea levels are likely to continue to rise for many years in the near future. Based on the studies of IPCC, it is clear that the climate change we face is because of sea-level rise. Further, the report reveals that the ocean and cryosphere both are in a changing climate state with global mean sea level, which is likely to rise between 0.29 and 1.1 m by the end of this century.

Radar altimetry has been tracking the changing height of the ocean surface and shows that global mean sea level has risen at the rate of over 3 mm every year and the rate of rise has increased in recent years. However, as with any average, the term "global mean sea level rise" does not tell the whole story. Sea level is not rising at the same rate everywhere. Throughout the globe, the coastal boundary may change and the ocean may recapture some parts of the carbon due to the very complex relationship between cryosphere–ocean and atmosphere.

Another satellite Jason data combined with ESA's Envisat, Cryosat, and the Copernicus Sentinel-3 satellites have provided essential data to monitor sea-level rise in the open ocean, precise mapping of rise nearer the coastline but it is more difficult. The reason behind this fact is that mountains, bays, and offshore islands distort the radar signal that is reflected back to the satellite. Besides the existing global warming, another problem is sea ice, which covers parts of the oceans in winter and is impenetrable to radar.

Through ESA's Earth Observation Science for Society Baltic Sea Level project, it was found that those sea-level changes in coastal areas were linked to sea glacier melting.

Data between 1995 and 2019 from satellite missions including the Jason series, Envisat, CryoSat, and Copernicus Sentinel-3, the team developed a multistage process. It was calibrated that with measurements from the various satellite missions using specially developed algorithms, it is possible to detect signals from the ice-covered seawater in the radar reflections produced along cracks and fissures called leads. It was possible to determine sea levels for the winter months, which also achieved better resolution of radar echoes close to land. It is now possible to measure sea level in coastal areas and compare the results with local tidal records. Sea level has risen at an annual rate of 2–3 mm in the south, which is less than in the Bay of Bothnia. The cause of this large rise is strong southwesterly winds that drive the waters to the north and eastward. This above-average increase in sea level does not pose a threat to coastal dwellers, however, because since the end of the last Ice Age, the land here has been rising by up to 1 cm a year as a result of post-glacial rebound.

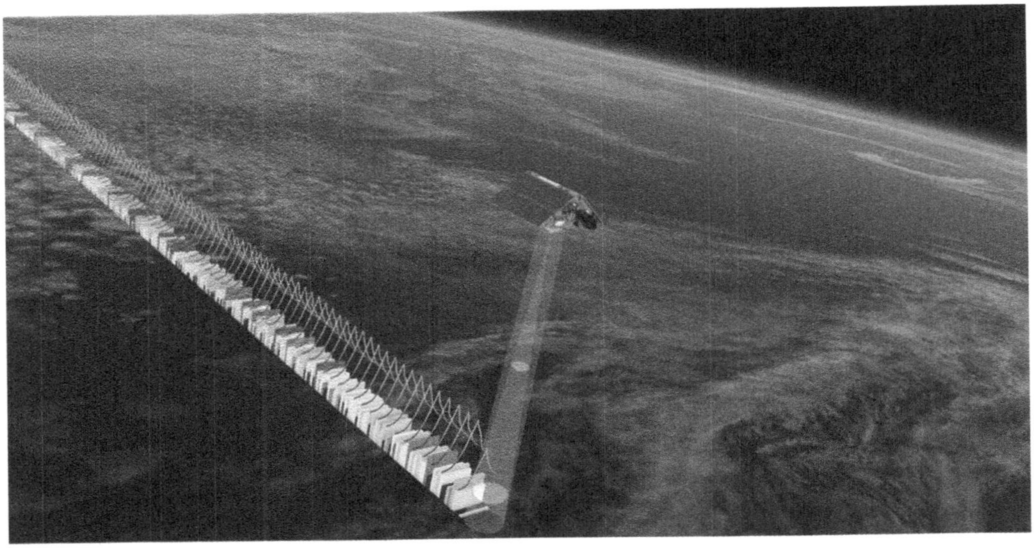

Figure 1.1 Copernicus Sentinel-6 radar altimeter. Source: ESA/ATG.

The researchers have also created a comprehensive dataset for the North Sea region. The sea level here is rising by 2.6 mm yr^{-1}, and by 3.2 mm in the German Bight. Local trends can be determined using the dataset and the user manual – both of which are freely accessible online (Figure 1.1).

The Baltic SEAL project has produced excellent results, demonstrating the value of Earth observation for society. This new way of processing satellite radar altimetry data is really going to help authorities to take appropriate measures to protect citizens and infrastructure from sea-level rise. We are also keenly awaiting the high-resolution data that will come from the new Copernicus Sentinel-6 mission, which also promises to map sea-surface height close to the coastline using an additional new processing algorithm called fully focused synthetic aperture radar to drastically increase the retrieved resolution (Courtesy: European Space Agency 2021).

The impact of the solar maximum can release more protons from the Sun during coronal mass ejection (CME) and continuous solar flares. This was evident in the power grid failures of Canada in 1959 and Northern India in 2012 and 2001. In all these incidents, the activity of the Sun showed sudden changes. The internal mechanism of the fluctuation of solar maximum and solar minimum is a complex subject that can be explained by the comprehensive study of the Forbursh decrease due to the sunspot starspot formation and impact of star burst on the magnetic field of the solar system. The effect of solar maximum may lead to the release of excessive ultraviolet rays from the Sun. CME from active Sun leads to more ultraviolet radiation. Ultraviolet radiation, along with its many other biomolecular effects, has an effect on actin protein. Lambda max of protein is 280 nm, which falls in the UV-B range. As a result of exposure to UV-B radiation of the same dose, which occurs during days of earth-directed CME to the actin monomer, the rate of formation of polymer from actin monomer is found to be lower. Actin polymer acts as a platform for various vital cellular reactions. Since the location of sunspots, which leads to CME, corresponds to a particular geo-coordinate of Earth, the entire biomass (as all living cells contain actin protein) of that particular area of the Earth will be affected adversely due to ultraviolet radiation (Geophysical Research Abstracts, Vol. 7, 00139, 2005 SRef-ID: 1607-7962/gra/EGU05-A-00139 © European Geosciences Union 2005).

Study of different planets of the solar system is important, keeping in view developing and understanding natural resources and natural disasters. The influence of extraterrestrial changes has different if any impacts on planets, which depend on the outer and inner composition of these celestial objects and their environment. Detailed discussion of mineralogical and morphological variability is the key to understanding the tectonic framework of the Moon, which has changed the concept of global information system. Keeping in view the terrestrial analog correlation, the study of the Moon and Mars is important. Using indigenous satellite Chandrayaan (Moon Satellite) and Mangalyan (Mars Orbiter Probe) when correlated with tectonic framework and mineral distribution on the surface, it has the potential to solve issues of sustainable development.

We have used data from the Mars Color Camera (MCC) and Thermal Infrared Spectrometer (TIS) payloads of the Mars Orbiter Mission. MCC returns "True Color" images in different resolutions. Hence, it is possible to provide high-resolution localized scenes (attaining a spatial resolution of 19.5 m pix^{-1} @ at periapsis) as well as a synoptic view of the planet depending upon the orbit. However, the coverage of high-resolution images was very low and often a shift in the geo-referencing was observed. TIS dataset was noisy and smoothening of the results often led to deviation from the actual observations. True color images of the MCC was an ingenious idea and we would recommend the same with perhaps better resolution in upcoming missions.

Mineralogy is an important aspect in studying geological evolution of a planet. Currently, OMEGA and CRISM sensors provide data in the near-infrared range (NIR) for studying secondary hydrated minerals. However, usage of OMEGA is not as widely popular as CRISM, probably because of the ease (also availability) in data processing. Nevertheless, current coverage of CRISM is reduced to half resolution short (HRS). Recent reports of artifacts in the dataset as well as the detector coolant malfunction (of CRISM) have raised questions of previous investigations, especially of sulfates. Thus, sending an Indian satellite to Mars with an IR (NIR + TIR) sensor which makes use of these research gaps would be very useful for scientists worldwide. At present we do not have a full Mars digital elevation model beyond 100 m (created by merging HRSC and MOLA datasets). But this dataset is also interpolated due to missing data in some areas. Therefore, we can work towards developing higher resolution altimeter as in the Chandrayaan mission.

Net Mars mission should be landing site inspired. Nevertheless, as the main focus of upcoming missions has shifted to finding biosignatures, it is important to first create a team of scientists with expertise in different areas to carry out field studies. This would help in understanding the kind of biosignatures that can be expected with least possible destruction due to heat in experiments, UV, etc. Based on liability of the compounds, we can focus on devising a sensor for specific classes such as lipids, proteins, amino acids, etc. A miniature XRF/MS/Raman spectrometer may help in detection of biosignatures intercalated within the rocks. Perhaps a subsurface exploration in this case may be a more fruitful and novel approach as other missions are planning to explore the surface. Recent results from Insight lander have opened a new area for exploration on Mars. Thus, magnetic and seismic studies of Martian subsurface (especially in the volcanically active areas of the past or deep crater basins) may give interesting results. Extraterrestrial remote sensing not only has the potential to infer the materialistic information of natural hazards and natural resources of the Earth and planetary system but can unveil the mystical universe.

Further Reading

Gomes, J.F.P., Mukherjee, S., Radovanović, M.M. et al. (2012). Possible Impact of the Astronomical Aspects on the Violent Cyclonic Motions in the Earth's Atmosphere. In: *Solar Wind: Emission, Technologies and Impacts*. Nova Science Publishers. ISBN: 978-1-62081-984-5 https://www.novapublishers.com/web/web_files/NewTitles/08_2012_August%20New%20Titles.pdf.

Mukherjee, S. (2003). Security Implications of Climate Change. In: *India's National Security – Annual Review* (ed. S. Kumar). India Research Press. ISBN 81-87943-56-4. http://www.indiaresearchpress.com.

Mukherjee, S. (2004). *Text Book of Environmental Remote Sensing*. New Delhi: Macmillan India Limited. ISBN: 1403922357. INBK103842.

Mukherjee, S. (2013). *Extraterrestrial Influence on Climate Change*. Springer. ISBN: 978-81-322-07290.

Mukherjee, S. (2020). *Effect of Star-Burst on Sun-Earth Environment Vision and Voyages for Planetary Science in the Decade 2013-2022*. USA: The National Academic Press https://solarsystem.nasa.gov/studies/41/effect-of-star-burst-on-sun-earth-environment/ (accessed 1 December 2017). https://doi.org/10.17226/13117 ISBNs: Paperback: 978-0-309-22464-2 E book: 978-0-309-20957-1.

2

Principles of Remote Sensing of Sun–Earth Climate

2.1 Introduction

There has been phenomenal growth in the field of remote sensing of Sun–Earth climate in the last two decades. In the late nineties, the concept of the influence of galactic cosmic rays on the Sun and the Earth was getting very little recognition; however, climate change is possible to explain with a number of extraterrestrial phenomena and changes. Extraterrestrial remote sensing infers that the Sun of the Earth is an unstable star and we have to live with its heliophysical variables. The Living with a Star (LWS) program of NASA also maintains the Targeted Research and Technology program.

The LWS program was initiated in 2000 and established as a new exciting research opportunity in 2001. NASA sponsored LWS, initiated by the Interagency Consultative Group; ILWS was formed in 2002. The program is focused on understanding the relationship between the space weather and environment of the Earth, and other planetary bodies comes under the purview of space science. Following are the focus of the study:

- Earth and planetary science
- Human exploration and development in planets of solar system
- Aeronautics and space transportation
  Currently the three active spacecraft dedicated to LWS program are Solar Dynamics Observatory launched in 2010, Parker Solar Probe in 2018, and the Solar Orbiter launched in 2020. In addition, the BARREL and Van Allen Probes are important extraterrestrial satellites that measure the fluctuation of Van Allen's Belt of Sun–Earth environment (Figure 2.1).

The Van Allen Probes (VAP), which measure geomagnetic field fluctuation and the possibility of trapping protons from the solar flare or sunspots, formerly known as the Radiation Belt Storm Probes (RBSP), were two robotic spacecrafts. These were used to study the Van Allen radiation belts that surround Earth. NASA conducted the VAP mission, which was a part of the LWS program. Understanding the radiation belt environment and its variability has practical applications in the areas of spacecraft operations, spacecraft system design, mission planning, and astronaut safety. The probes worked from 2012 August to 2019 December. The data of this radiation belt shows some fluctuation, which was correlated with various other extraterrestrial data including cosmic rays from different countries. It shows that solar cosmic rays are sympathetic with heliophysical parameters; however, galactic cosmic ray shows anticorrelation with proton flux, electron flux, and planetary indices. The cosmic ray detector installed in Jawaharlal Nehru University measures cosmic rays per unit time from the Sun and galactic sources. This instrument measures space environment in the form of solar and galactic cosmic rays. Rise in galactic cosmic rays has shown

Extraterrestrial Remote Sensing and Climate Change, First Edition. Saumitra Mukherjee.

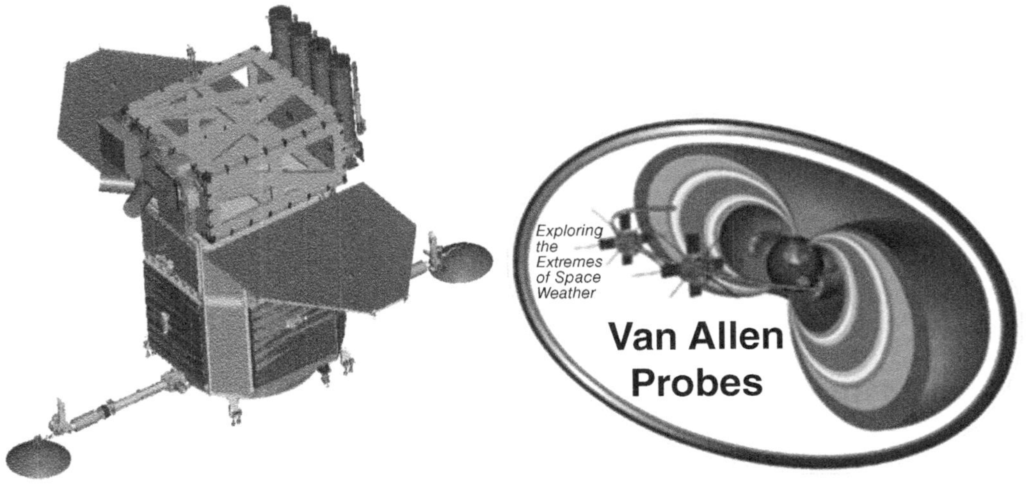

Figure 2.1 Solar Dynamics Observatory and Van Allen Probe.

reduction in heliophysical parameters including electron flux, proton flux, and planetary indices. It has been observed that the heliophysical as well as galactic cosmic ray intensity changes can influence the environment of the planetary system including the atmosphere of the Earth. Thunderstorms in Delhi, India, have shown the correlation with the cosmic ray and heliophysical parameters. Terrestrial environment change manifestations such as vegetation vigor can be measured by the satellite data, which also has positive correlation with the solar flux.

Climate disturbances are being studied and monitored as a result. Some of its factors are still unanswered. A correlation is being attempted to understand Earth-directed coronal mass ejection (CME), increase in electron flux, and magnetic indices (Kp values). Catastrophic weather and climatic change considers the violent meteorological events that seem to be occurring with greater frequency. For example, why do hurricanes always move from east to west? There is a possible correlation between the number of hurricanes that develop each year and average sea surface temperatures. Perhaps the greatest question regarding fierce weather events concerns what they might tell us regarding changes in our climate. How do these occurrences and such global phenomena as El Niño and the attendant "strange weather" fit into present models of climatic change? CME is being correlated with the change in Tropical Ocean–Atmosphere Interactions. Most of the time Earth-directed CMEs produce auroras, which are visible in Polar regions of the Earth. During the Earth-directed CME, it has been observed that the sudden change in atmosphere followed sudden weather change. Intermittent Earth-directed CMEs were observed with the fluctuation in solar flux. It has been observed that sudden decrease in Kp values in a geo-specific sunspot influences low electron flux. This phenomenon leads to lowering of atmospheric temperature locally, which leads to rainfall (if clouds are present in surrounding) or snowfall in higher altitude and latitudes. It can be suggestive that the material ejected from the sunspots as CME is in fifth state of matter (not plasma). The influence of this material reaches Earth within 36 hours of its ejection from the active sun spot.

*The postulated 5th State of Matter and its Influence on Climate Change was presented in European Geosciences Union in 2005 by the author and L. Kortvelyessy. Geophysical Research Abstracts, Vol. 7, 00138, 2005 SRef-ID: 1607-7962/gra/EGU05-A-00138 © European Geosciences Union 2005.

Measurement of cosmic rays in different parts of the world has revealed that it varies in its characteristics as well as intensity in space and time. Sun and Heliophysical Observatory (SOHO) satellite data reveals heliophysical parameter changes before sudden changes between the Sun and the Earth. The influence of extragalactic changes on the Sun as well as the Sun–Earth environment seems to be both periodic and episodic [1]. The periodic changes in terms of solar maxima and minima occur every six years, whereas the episodic changes may occur suddenly. These changes can be monitored as a sudden increase or decrease by cosmic ray detectors. Heliophysical variables and cosmic ray intensity on the environment of the Earth can be studied together. It has been observed that the atmospheric electric field changes with the fluctuation of the cosmic rays from galactic sources. Throughout the world, the phenomena of thunderstorms and cloudbursts are increasing. Cosmic ray variability has shown a positive correlation with climate change. Extraterrestrial and solar influence on climate change not only generates erratic rainfall and inconsistent atmospheric temperature, but it has major impact on the vegetation also.

Extragalactic cosmic rays represent an energy spectrum, which is being received by the solar system. The particles of cosmic rays are atom–nuclei with almost light velocity. The effects of the Sun on the environment of the Earth were found to be modulated by the geomagnetic field and the ionizing potential of the cosmic rays. Earth-directed CME and its effects on the thermosphere, ionosphere, and atmosphere were studied. During Earth-directed CME, a beam of electrons (plasma) is pumped toward Earth. This beam of electrons is highly conductive. It generates an electric field that is transmitted to Earth's natural plasmosphere and ionosphere. This thin layer of charged electric field further influences the ionosphere and atmosphere of Earth. Since a beam of electrons carry an electric current, a magnetic disturbance would be produced. The SOHO satellite provides heliophysical data including electron flux (E-flux), proton flux (P-flux), and planetary indices (Kp). These data are continuously updated by NASA and being made available through the website www.spaceweather.com

2.2 SEVAN Data Interpretation for Solar Influence on Climate Change

Space Environment Viewing and Analysis Network (SEVAN) is recording the cosmic ray and atmospheric pressure at Jawaharlal Nehru University, India. The cosmic rays have been recorded continuously since December 2010 to infer the correlation of the atmospheric pressure with the cosmic ray intensity from the galactic sources and from the Sun. A number of different environmental events have been recorded in which the atmospheric changes have been correlated with cosmic rays from the Sun.

The heliophysical as well as cosmic ray data are available for any user across the world.

Similar SEVAN units are functioning in other parts of the world, which has also shown the similar phenomena of the selective rise and fall of cosmic rays from different sources before thunderstorms.

Besides these, the influence of solar flux has been recorded by using solar flux meter. The data shows variations with the spectral response from the vegetation, which has been recorded by using LANDSAT satellite data of the same dates.

On 30 May 2014, a sudden thunderstorm in Delhi, India, was unprecedented. The wind speed was at the rate of 92–100 km h^{-1}, which claimed 23 human lives, uprooted thousands of trees, and collapsed several buildings. Routine weather prediction of different agencies was not able to predict this catastrophe (Figure 2.2).

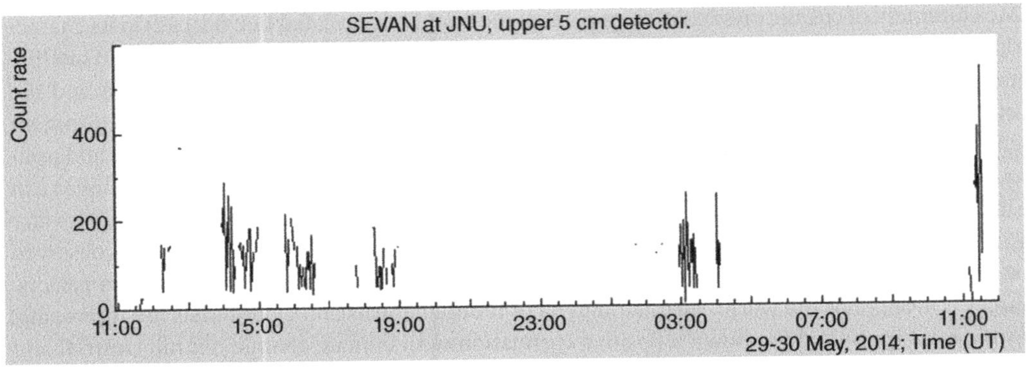

Figure 2.2 Upper 5 cm scintillators of SEVAN detector are capable of recording galactic cosmic rays. The Jawaharlal Nehru University, India, device has recorded sudden rise in the cosmic ray intensity from 11.40 a.m. to 11.45 p.m. on 29–30 May 2014 before the catastrophic thunderstorm in Delhi. Source: Data from PTI 2014.

The SEVAN detector has multiple sets of scintillators to record the incoming cosmic rays of different origin. The topmost 5 cm thick scintillator has the potential to record the galactic cosmic rays, but the middle unit of 25 cm thickness can record the cosmic rays from the Sun. It has been recorded that six hours before the thunderstorm in Delhi on 30 May 2014, the topmost unit of SEVAN recorded a sharp rise in cosmic rays, while the middle unit of the device recorded low cosmic ray intensity than the normal. The pressure sensor of SEVAN shows a sympathetic rise with the 5 cm cosmic ray sensor (Figures 2.3 and 2.4).

Terrestrial satellite data shows the recording of spectral response in different parts of the electromagnetic spectrum including visible and near-infrared. An attempt has been made by using the LANDSAT satellite data to generate normalized difference vegetation index (NDVI) in different

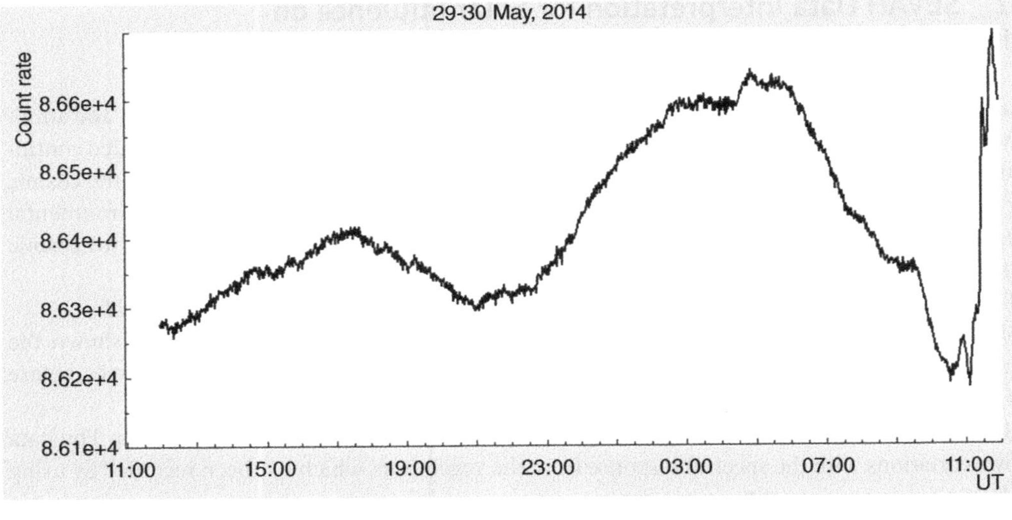

Figure 2.3 Pressure sensor of SEVAN JNU showing sudden rise in atmospheric pressure six hours before the thunderstorm. Source: Data from SCACRD http://www.crdfriends.org/.

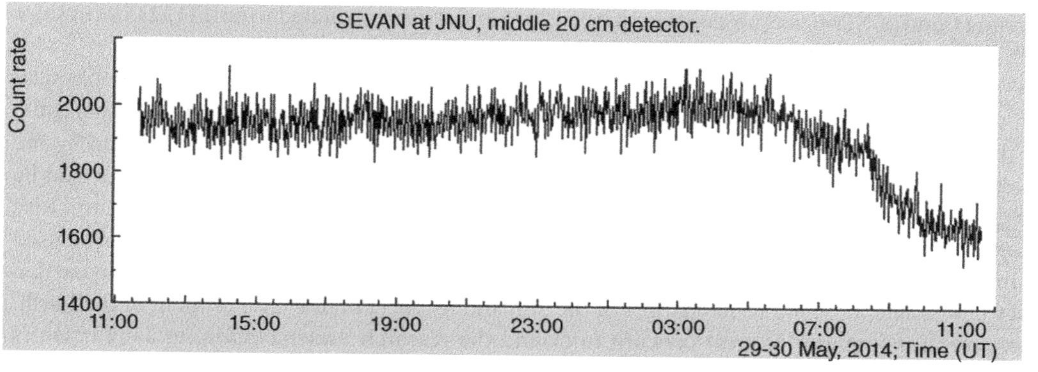

Figure 2.4 Twenty-five-centimeter thick cosmic ray sensor of SEVAN JNU detects solar particle flux showing sudden fall in count rate six hours before the thunderstorm. Source: Data from SCACRD http://www.crdfriends.org/.

parts of Delhi, India. The NDVI value has been generated using ERDAS digital image processing software with the following algorithm:

NDVI = Reflected Near Infrared/Reflected Red/Reflected Near Infra Red + Reflected Red

NDVI value shows the vegetation vigor, which may vary in space and time. This value ranges from +1 to −1. It has been inferred that the solar lux is showing sympathetic relation with the NDVI (Figure 2.5).

On 27 January 2015, the NDVI value was 0.0684 with solar flux reading 33.34111 k lux, while on 19 February 2015, the NDVI value was 0.085 with corresponding solar flux of 46.95793 k lux.

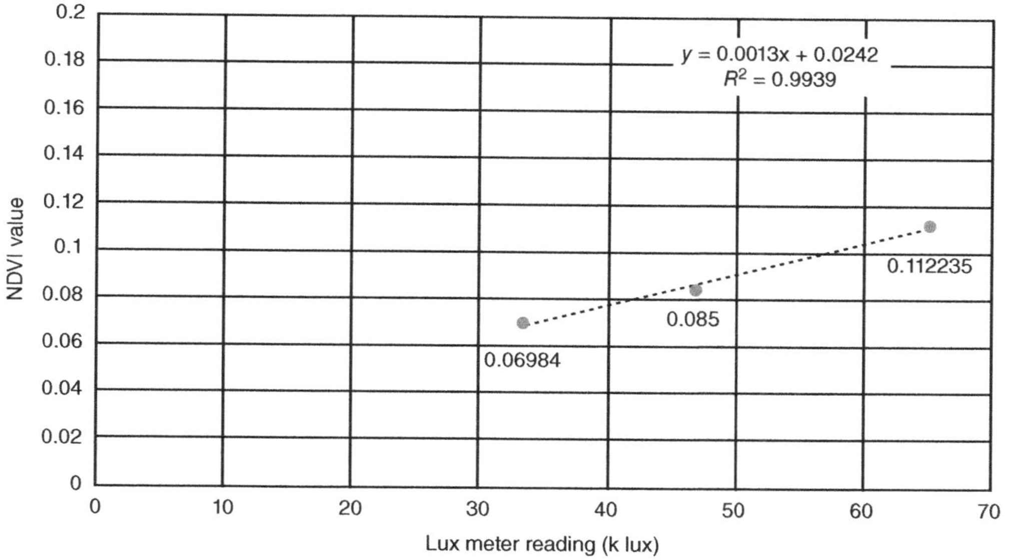

Figure 2.5 Rise in NDVI value on different dates at the Jawaharlal Nehru University, Delhi, India, shows sympathetic relation with the solar flux recorded by the lux meter. Source: Data from SCACRD http://www.crdfriends.org/.

A rising trend of NDVI was observed on 8 April 2015 and it was calculated to be 0.112235 with solar flux of 65.27059 k lux.

Cosmic rays from the Sun and galactic sources when correlated with the changes in heliophysical parameters give a possible relation. Terrestrial satellite data shows a correlation with the solar flux. In the realm of climate change, the prediction of environmental changes can be boosted by the study of the variations in atmospheric temperature and vegetation health. This technique can be used successfully for thunderstorm prediction followed by rise in atmospheric temperature. Rise in atmospheric temperature with high solar flux and elevated heliophysical parameter can be used to infer the health of the plants by rising trend in NDVI value. Extensive work has been carried out in the concept of long-term changes of the Sun and its effect on the environment of the Earth. However, it is a matter of great concern to extend the research understanding to daily changes inside the Sun and its effect on the environment.

2.3 Changes Within the Sun in 24 Hours and Its Effect on the Environment of the Earth

The solar illuminations are subject to daily and seasonal changes. Every day, the sun rises in the east, assumes a high position in the sky at noon, and sets in the west. In addition to the daily changes, there are also seasonal changes that depend to a large extent on the latitude of the observer. For example, in Central Europe the time between sunrise and sunset can vary by almost eight hours. Just as noticeable are the different heights above the horizon at noon in summer and in winter.

Sun varies in every way we can observe it. The magnetic field of the Sun varies by the movement of plasma inside it. The plasma is a charged colloidal material with dipole, which changes on the time scale from millisecond to billions of years. The magnetic field generated by sunspots and solar flares normally reaches its peak every 11 years but if the galactic changes are abnormally higher, then this cyclic movement of the sunspot cycle is deferred. The constellation of the stars, which also produce magnetic field, can produce cosmic rays, which modulate the effects of the Sun (Figure 2.6).

A correlation has been done in extraterrestrial remote sensing at the Laboratory of School of Environmental Sciences, Jawaharlal Nehru University, Delhi, India. It infers that on those dates when the NDVI of Delhi and surrounding areas (56 locations) (Table 2.1) is greater, the solar activity is also greater; however, the SEVAN node of JNU shows a low value (Figure 2.7).

It has been well documented that climatic change has the potential to influence the environment of the Earth in space and time. Influence of heliophysical parameters and galactic cosmic rays on the environment and health of living beings on Earth are being studied (Figure 2.8). The outbreak of pandemic COVID-19 and the future threat of various health hazards for the living organisms of the Earth need to be studied in detail by the variables of sunspots, star spots, and cosmic ray data. The shift of the Earth's magnetic pole toward Siberia is a phenomenon of change in the Earth's interior as well, triggered by changes beyond the Earth. The manifestation of polar wandering is climate change due to solar and galactic changes. Polar wandering, terrestrial climate change, and the COVID-19 pandemic outbreak are linked with each other. It is a series of events initiated in outer space in the form of changes in the galactic plane. Impact of the galactic plane has changed the Sun–Earth–Cosmic equilibrium. More galactic cosmic rays and fewer ultraviolet rays from the Sun due to improved ozone layer have changed the mutation probability of viruses. Furthermore, the persistent solar minimum of 2019–2020 is a near replica of the 2009–2010 solar minimum.

Figure 2.6 Lux meter measuring the daily fluctuation of the Sun.

Table 2.1 NDVI values showing sympathetic relation with cosmic ray and solar flux.

Date of satellite image (LANDSAT)	27 January 2015	19 February 2015	23 March 2015	8 April 2015
NDVI	0.33	0.46	0.46	0.65
Average K-Lux	0.06	0.085	0.85	0.19
Cosmic Ray count per second (JNU data)	100	12	500	22

This extraterrestrial change may affect terrestrial environment and health of the living beings of the Earth by evolution of new species of virus. The mechanisms of virus mutations are still not very clear. However, the sudden outbreak of pandemics like the swine flu of 2009 and COVID-19 of 2019–2020 has compelled us to see the triggering of this catastrophe beyond the Earth. A possible correlation can be initiated based on the solar-minimum-induced climate change and its implications on triggering viral mutations. Variations in star spots can control sunspots and its associated heliophysical parameters. The wandering of Earth's north magnetic pole, the location where the magnetic field points vertically downwards, has long been a topic of scientific fascination. Since the first *in situ* measurements in 1831 of its location in the Canadian arctic, the pole has drifted inexorably toward Siberia, accelerating between 1990 and 2005 from its historic speed of 0–15 km/yr to its present speed of 50–60 km/yr. In late October 2017, the north magnetic pole crossed the

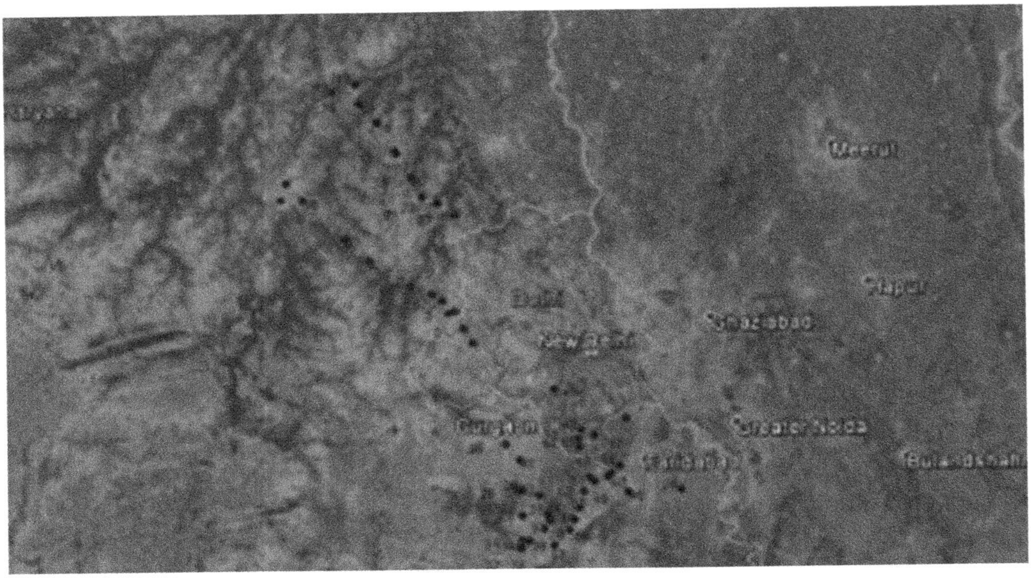

Figure 2.7 Black dots in Delhi and surrounding areas are showing locations of solar flux and NDVI calculation points.

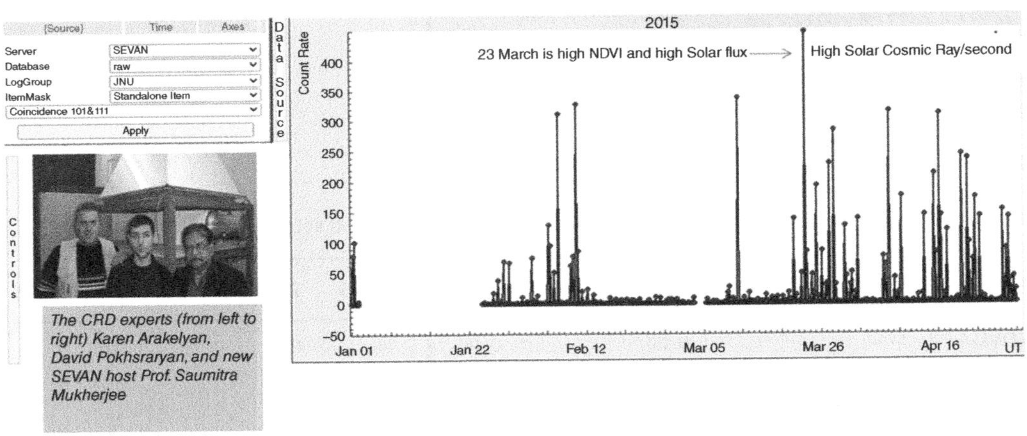

Figure 2.8 Solar cosmic ray rise showing positive correlation with NDVI and solar flux.

International Date Line, passing within 390 km of the geographic pole, and is now moving southward. Here we show that over the last two decades, the position of the north magnetic pole has been largely determined by two large-scale lobes of negative magnetic flux on the core–mantle boundary under Canada and Siberia. Localized modeling shows that elongation of the Canadian lobe, probably caused by an alteration in the pattern of core flow between 1970 and 1999, substantially weakened its signature on Earth's surface, causing the pole to accelerate toward Siberia. A range of simple models that capture this process indicate that over the next decade, the north magnetic pole will continue on its current trajectory, traveling an additional 390–660 km toward Siberia. Based on the observation of extraterrestrial changes, the environmental changes of the Earth are well established. It is essential to infer the periodic and sudden changes in the environment of the Earth and try to correlate it with the heliophysical and galactic cosmic ray data. If the global environment is monitored with extraterrestrial changes, then it will be possible to establish the natural hazard warning system. Heliophysical parameters can be downloaded from the space weather website, and cosmic ray per unit time can be detected globally from the following stations:

Particle detectors
AMMM – **A**ragats **M**ultichannel **M**uon **M**onitor
ArNM – **A**ragats **N**eutron **M**onitor
ASNT – **A**ragats **S**olar **N**eutron **T**elescope
Cube – Neutral Cosmic Ray Monitor
Cube_3 cm – Cube Monitor
MAKET – MAKET-ANI Extensive Air Shower Detector
NaI – NaI(Tl) Monitor
NAMMM – **N**or-**A**mberd **M**ultidirectional **M**uon **M**onitor
NANM – **N**or-**A**mberd **N**eutron **M**onitor
SEVAN – SEVAN Monitors
Stand 1 cm – Plastic Scintillators Monitor for Low Energy Particles
Stand 3 cm – Plastic Scintillators Monitor with Spectral Analysis

SEVAN Monitors
SEVAN Aragats – SEVAN Monitor at Aragats
SEVAN Nor-Amberd – SEVAN Monitor at Nor-Amberd
SEVAN Yerevan – SEVAN Monitor at Yerevan
SEVAN Byurakan – SEVAN Monitor at Byurakan
SEVAN Moussala – SEVAN Monitor at Moussala
SEVAN Zagreb – SEVAN Monitor at Zagreb
SEVAN JNU – SEVAN Monitor at Jawaharlal Nehru University, Delhi, India

Other monitors
Electric Field – Electric Field Monitor
Lightning Detector – Lightning Detector
LEMI-417 – Magnetotelluric Station (MTS) at Aragats
LEMI-018 – Magnetotelluric Station (MTS) at Nor Amberd
Weather Station – Davis Wireless Vantage Pro2 Plus

Reference

1 Shaviv, N.J. and Veizer, J. (2003). Celestial driver of phanerozoic climate. *GSA Today* 13 (7): 4–10.

Further Reading

Dorman, L.I. and Dorman, I.V. (2005). Possible influence of cosmic rays on climate through thunderstorm clouds. *Adv. Space. Res.* 35: 476–483.

Frigo, E., Pacca, I.G., Pereira-Filho, A.J. et al. (2013). Evidence for cosmic ray modulation in temperature records. *Ann. Geophys.* 31: 1833–1841.

Midya, S.K. and Gole, P.K. (2013). Trend of major earthquake during 1900-2011 and its association with some solar and geomagnetic parameters. *Ind. J. Phys.* 88 (1): 1–4.

Mukherjee, S. (2007a). Changes in heliophysical parameter influence on environment of the earth. *Bull. Astr. Soc. India* 35: 1–7.

Mukherjee, S. (2007b). Proceeding International Heliophysical Year-Coordinated Investigator Programme, Physik-Zentrum, Bad Honnef, Germany (18–19 May 2007). http://www.ieap.uni-kiel.de/et/ag-heber/ihy2007/sun-heliosphere-earth/cip/program/index.php (accessed 19 May 2021).

Mukherjee, S. (2015). Rise in galactic cosmic rays before thunderstorm in Delhi. *J. Phys. Maths* 6 (131): https://doi.org/10.4172/2190-0902.1000131.

Mukherjee, S. (2021). Changes in heliophysical parameters on global environment and health. Heliophysics 2050 White Papers 2021. https://www.hou.usra.edu/meetings/helio2050/pdf/4131.pdf https://www.hou.usra.edu/meetings/helio2050/pdf/helio2050whitepapers_program.htm (accessed 20 January 2021).

Shibita, S., Munakata, K., Yasues, S. et al. (2004). Effects of atmospheric electric fields on cosmic rays. *Phys. Rev.* 69: 123010.

Shyam, A. and Kaushik, T.C. (1999). Observation of neutron bursts associated with atmospheric lightning discharge. *J. Geophys. Res.* 104 (A4): 6867–6686.

Sikka, P., Selvam, A.M., Ramachandra, A.S and Murty, R.(1998). Possible solar influence on atmospheric electric field. ArXivi: chao-dyn/9806014vi.

Venkatesan, D. and Badruddin (1989). Cosmic ray intensity variation in the 3 dimensional variations. *Space Sci. Rev.* 52: 121–194, 10.10.107/BF00704241.

Walker, B. and Steffen, W. (1997). An overview of the implications of global change for natural and managed terrestrial ecosystems. *Conservation Ecology (online)* 1 (2): 2. http://www.consecol.org/voll/iss2/art2.

3

Extraterrestrial Remote Sensing Satellites

3.1 Introduction

The activities and characteristics governed by the solar dynamo Sun are very complicated. The Sun pulsates in an 11-year interval. Solar wind, corona, and flares show periodicity in propagating its heliophysical parameters. YOKOH, SOHO, and ULYSAS are the names of a few solar monitoring satellites.

Solar and Heliospheric Observatory (SOHO) is a prestigious science project of international cooperation between the European Space Agency and NASA to study the Sun, from its deep core to the outer corona, and the solar wind. Together with ESA's Cluster mission, SOHO is studying the Sun–Earth interaction from different perspectives. SOHO's easily accessible, spectacular data, and basic science results have captured the imagination of the space science community and the general public alike. Space weather monitoring satellite SOHO was designed to address mainly some pertinent unsolved issues about the Sun–Earth Cosmic. Its focus is mainly about the structure and dynamics of the solar interior mechanism to generate hotspots, which develop as sunspots after emission of plasma from the solar corona. During the solar minimum, sudden rise in solar flare leading to extreme solar wind production is still a puzzle. Helioseismic evolution and its manifestation as seismic waves produced in the turbulent outer shell of the Sun appear as ripples on its surface. Science highlights to date from SOHO have provided an unprecedented breadth and depth of information about the Sun, from its interior, through the hot and dynamic atmosphere, to the solar wind, and its interaction with the interstellar medium. These findings have been documented in an impressive, still growing body of scientific and popular literature. Some of the key results reveal the first image of Sun's convection zone and of the structure of sunspots below the surface. SOHO precisely measures the most detailed and precise measurements of the temperature structure, the interior rotation, and gas flows in the solar interior. In addition, it measures the acceleration of the slow and fast solar wind. It identifies the source regions and acceleration mechanism of the fast solar wind in the changing magnetic regions at the Sun's poles. New dynamic solar phenomena are also being monitored such as coronal waves and solar tornadoes. Thirty-six hours prior information about space weather monitoring is very important information, which gives up to three days of notice of Earth-directed disturbances and plays a lead role in the early warning system for space weather. Solar irradiance was considered to be a constant, but the SOHO observation has been monitoring the total solar irradiance in different time as well as variations in the extreme ultraviolet flux. However, it is important to understand the impact of solar variability on Earth's climate. Besides watching the Sun, SOHO has become the most prolific discoverer of comets in astronomical history: as of January 2011, more than 2,000 comets had been found by SOHO. Orbit SOHO moves around the Sun in step with the Earth, by slowly orbiting around

Extraterrestrial Remote Sensing and Climate Change, First Edition. Saumitra Mukherjee.

the First Lagrangian Point (L1), where the combined gravity of the Earth and Sun keep SOHO in an orbit locked to the Earth–Sun line. The L1 point is approximately 1.5 million kilometers away from Earth (about four times the distance of the Moon), in the direction of the Sun. There, SOHO enjoys an uninterrupted view of our daylight star. Mission lifetime SOHO was designed for a nominal mission lifetime of two years. Because of its spectacular successes, the mission was extended five times (in 1997, 2002, 2006, 2008, and 2010). This allowed SOHO to cover an entire 11-year solar cycle (#23) and the rise of the new cycle 24. SOHO is currently approved through the end of 2012. Control of the spacecraft was lost in June 1998 and only restored three months later through efforts of the SOHO recovery team. All 12 instruments were still usable, most with no ill effects. Two of the three onboard gyroscopes failed immediately and the third in December 1998. After that, new onboard software that no longer relies on gyroscopes was installed in February 1999. It allowed the spacecraft to return to full scientific operations while providing an even greater margin of safety for spacecraft operations. This made SOHO the first three-axis stabilized spacecraft operated without gyroscopes, breaking new ground for future spacecraft designs.

Solar maximum is repeated every 11-year time period. The plasma within the Sun is electrically charged colloidal material, which reaches from the core to the surface of the Sun in a 22-year period. The plasma in a single turn over within the Sun is polarized in each hemisphere in this 22-year cycle. During one 11-year solar maximum, the sunspot appears in the northern hemisphere while the second phase sunspots maximum appears in the southern hemisphere.

The dynamics of the Sun found out to be strongly affecting the interplanetary and terrestrial environment, causing phenomena that have a great impact on the man-made activities. Using various extraterrestrial satellites, the Earth's magnetosphere–ionosphere–geosphere–biosphere–hydrosphere system is being recorded by various satellites. Extraterrestrial Satellite SOHO and its sensors, which record the changes of the solar wind and interplanetary conditions due to the solar activity, has been widely investigated, showing how the dynamics of the coupled solar wind–magnetosphere–ionosphere (SMI) system resembles that of a complex system, showing scale-invariant features, turbulence, and a near-criticality behavior. On the other hand, in the framework of dynamical systems, several new tools and methods have been proposed to quantify and characterize the dynamical complexity and its role in nonlinear out-of-equilibrium dynamical systems.

The SOHO is a joint program of the European Space Agency and NASA to understand the Sun from its core to the periphery. This satellite was launched in 1995.

Following instruments are installed in SOHO:

CDS: Coronal Diagnostic Spectrometer
CELIAS: Charge, Element, and Isotope Analysis System
COSTEP: Comprehensive Superathermal and Energetic Particle Analyzer
EIT: Extreme Ultraviolet Imaging Telescope
ERNE: Energetic and Relativistic Nuclei and Electron Experiment
GOLF: Global Oscillations at Low Frequencies
LASCO: Large Angle and Spectrometric Coronagraph
MDI: Michelson Doppler Imager
SUMER: Solar Ultraviolet Measurements of Emitted Radiation
SWAN: Solar Wind Anisotropies
UVCS: Ultraviolet Coronagraph Spectrometer
VIRGO: Variability of Solar Irradiance and Gravity Oscillations

Large Angle Spectrometric Coronagraph (LASCO), which is able to take images of the solar corona by blocking the light coming directly from the Sun with an occulted disk, creates an artificial eclipse within the instrument itself. The position of the solar disk is identified in the images by the white circle. The most prominent feature of the corona is usually the coronal streamers, those nearly radial bands that can be seen as image (a) and (b) (Figure 3.1).

Solar prominence is a sudden outburst from the surface of the Sun, which has powerful magnetic loops and which influences Sun–Earth environment (Figure 3.2).

SOHO inferred helioseismology investigations primarily aiming at the study of those parts of the solar oscillations spectrum that cannot be obtained from the ground. The required sensitivity for observing the very low modes and the high modes is difficult to achieve from the ground because of noise effects introduced by the Earth's diurnal rotation for the low modes, and the transparency and seeing fluctuations of the Earth's atmosphere for the high modes. The solar atmosphere remote sensing investigations are carried out with a set of telescopes and spectrometers that will produce the data necessary to study the dynamic phenomena that take place in the

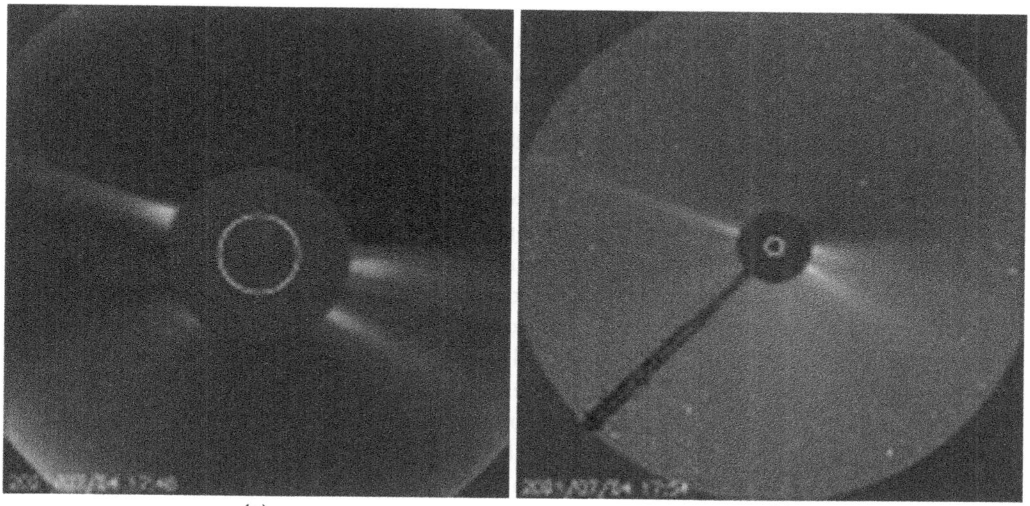

(a) (b)

Figure 3.1 Occasionally a coronal mass ejection can be seen being expelled away from the Sun and crossing the field of view of both coronagraphs. The shadow crossing from the lower left corner to the center of the image is the support for the occulted disk. Source: LASCO Image (Courtesy: ESA-NASA).

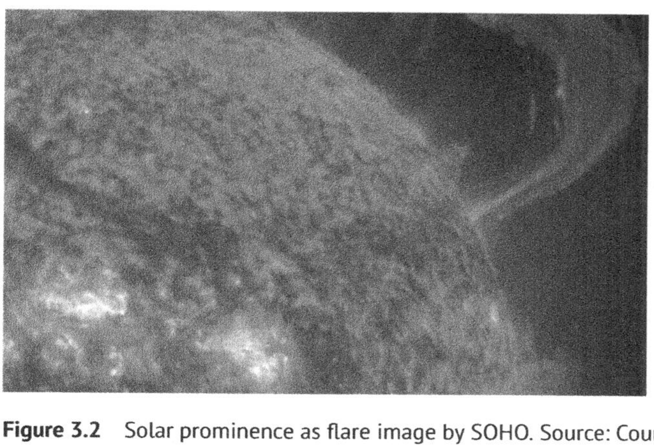

Figure 3.2 Solar prominence as flare image by SOHO. Source: Courtesy ESA-NASA.

solar atmosphere at and above the chromospheres. The plasma is being studied by spectroscopic measurements and high-resolution images at different levels of the solar atmosphere. Solar plasma diagnostics obtained with these instruments provide temperature, density, and velocity measurements of the material in the outer solar atmosphere.

3.2 Workstation Requirements

Three types of workstations are envisioned at the Experiment Operations Facility (EOF): the workstations of resident, and in some cases nonresident, Principal Investigator (PI) teams (Instrumental Workstation or IWS), the Project Scientists' workstation, and the Science Operations Coordinator's workstation. The Science Operations Coordinator's workstation is an EOF-resident workstation for use by SOHO science operations personnel, under the direction of the Science Operations Coordinator. This workstation will be used to receive and display data used in planning (e.g. from ground-based observatories) to resolve connects in instrument operations so as to produce weekly science operations schedules and maintain various databases (such as the key parameters) to be transmitted to Data Distribution Facility (DDF) and International Solar Terrestrial Physics (ISTP) Central Data Handling Facility (CDHF). The Project Scientists' workstation is an EOF-resident workstation especially for the use of the Project Scientists, but can also be used as a temporary backup to the Science Operations Coordinator's workstation in the event of unavailability. A PI team that has made arrangements with the Project Scientists to implement IWS software on the Project Scientists' workstation will have a similar backup capability in case of IWS failure. Nonresident investigators will coordinate their operational commanding activities via one of the EOF-resident workstations, normally the Science Operations Coordinator's workstation. Commands from remote sites are to be submitted in a format ready for variation and validation before they are relayed to the Soho Mission Operations Control Center (SMOCC). Software to perform such functions on the EOF-resident workstation is to be furnished by each PI team requiring this capability. All EOF workstations will support agreed-upon standard software (Motif or Motif-compatible implementations of the Open Look toolkit, IDL, SQL) and formats (FITS) to facilitate the exchange of catalog, scientific, engineering, and planning data. Security provisions to ensure that command transmissions from remote sites over public networks do in fact originate with the authorized PI team and have been received intact are to be installed on the remote and EOF-resident workstations. All EOF workstations will be compatible with standard US power (60 Hz, 110 V) and receptacles platforms, which are not capable of supporting a subset of the agreed standards (e.g. certain PC Unix implementations for which IDL is not available, but which can open Motif windows on other EOF systems with IDL). Each EOF-resident investigation team will provide a sufficient number of dedicated workstations to acquire incoming data, process and/or monitor those data as required, and prepare schedules and command sequences for submission to the SMOCC. Each EOF workstation will provide its own capabilities for interfacing with the GSFC EOF local area network (LAN), including provisions for any special additional data interfaces. Each EOF workstation will provide its own capability for storage of instrument data and resident databases, and for data display and hardcopy. Each EOF workstation will provide its own capability to obtain data from a variety of databases, with elements on a common server as well as on individual teams' workstations. The individual instrument databases should be accessible to other SOHO users at the EOF. Each PI team will develop and maintain a catalog of accessible data recorded in different time domain for correlation.

The operation of the coronal payload on SOHO will be performed in three layers. Standard operations will involve planning sessions at the EOF with targets and operating sequences for one or more days prior to the operation. This is adequate for most solar targets. The second layer involves developments in solar activity that may demand changes in operation overriding previous planning, and this can be done by commanding from the ground during real-time passes. For the shortest time-scale transient activity, such as the build-up of a bright point, the onset of a flare, or an eruption, the EOF real-time interruption is not quick enough. Therefore, the third layer of operation requires the use of an inter-instrument flag.

3.3 Sun Heliophysical Observatory (SOHO) Interdisciplinary Science Matrix

The goals of SOHO require comparison and analysis of datasets from very different experiments. Such interdisciplinary studies require careful planning, prior to the observations, and involve complex analysis procedures. In recognition of this, the SOHO Coronal and Particle Working Group (SCPWG, now merged to the Science Planning Working Group, SPWG) initiated a study to provide an overview of the nature of such activities. The underlying goal of the ESA Solar Terrestrial Science Programme is to develop an understanding of the activity of the Sun and its influence on the Earth. SOHO and the Cluster spacecraft were designed to provide the backbone of such a study. Even with such a coordinated effort, it is difficult to correlate solar and space plasma events and structures. Complexities arise because of time delays and the uncertainty of propagation paths, as well as the linking of observations made with fundamentally different instruments, such as spectrometers and particle detectors.

At Jawaharlal Nehru University School of Environmental Sciences, the extraterrestrial remote sensing data of SOHO is being correlated with SEVAN detector; it is a node of Space Environment Viewing and Analysis Network developed with the funding of NASA, USA, and technical support of CRD, Armenia.

A new Constellation of Satellites from the European Space Agency can detect minuscule changes in the magnetic field of planets (Figure 3.3).

Before the occurrence of earthquake several times, it has been observed that the magnetic field between Sun and the Earth changes. The time-series analysis shows that Earth's magnetic field is decaying up to 5% in some places. These places where the magnetic field decreases galactic cosmic rays increases. It may be a far-reaching consequence but the magnetotail has been mapped by satellite before occurrence earthquake in Gujarat, India, on 26 January 2001.

Detailed investigation of magnetotail can be possible by interpretation of data of the extraterrestrial satellite geotail. This satellite was launched by NASA (USA) and ISAS (Japan) in 1992, which provides response of Earth's magnetic field from the extraterrestrial effect like coronal mass ejection from the Sun (Figure 3.4).

The constellation of Geotail Satellite has sensors, which measure following parameters:

1. Electrical Field Detector
2. Magnetic Field Measurement
3. High Energy Particle experiment
4. Low Energy Particle Experiment
5. Plasma Waves Investigation
6. Energetic Particle and Ion Composition Experiment
7. Comprehensive Plasma investigation

Electric field detector: The electric field detector studies the coupling of the E-field in the near-Earth magnetosphere and in the ionosphere (in particular during sub storms). The electric field antenna is coupled to an electron beam device with a range of 0.01–10 mV m^{-1}. Part of a package of

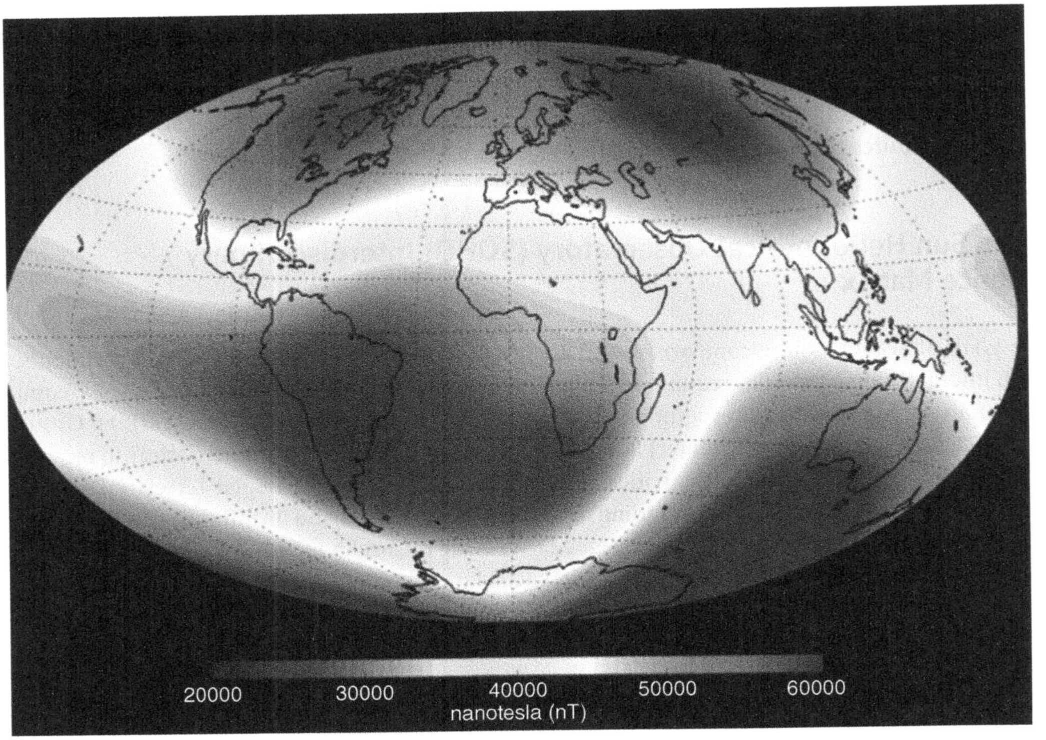

Figure 3.3 Magnetic field at the Earth's surface measured in Nano Tesla. Source: ESA.

Figure 3.4 Geotail Satellite. Source: Courtesy NASA-ISAS.

instruments observes the magnetosphere by navigating its tail. The scanning technique is traveling across the magnetosphere by navigating its tail. Its temporal resolution is excellent, full magnetospheric volume; sampling is done at 0.016 seconds interval.

Magnetic field measurement (MGF): This sensor is used to study the transport dynamics of mass, momentum, and energy between the magnetospheric and ionospheric plasma and to study the merging into the magnetotail. MGF is a fluxgate and search coil magnetometer for DC and AC measurements. Two sets of fluxgate magnetometers were deployed at a distance of 4 and 6 m mast. It measures with an accuracy of 0.006 nanotesla.

High-level particle experiment (HEP): It measures high-energy particles traveling through the magnetosphere up to 25 MeV for electrons, 35 MeV for Protons, and 210 MeV $charge^{-1}$ for ions. It has an assemblage of five instruments with a low-energy particle detector, a burst detector, two medium energy isotope detectors, and a high-energy isotope detector. Energy ranges from 0.02 to 2.5 MeV (electrons), 0.002 to 35 MeV (protons), 0.004 to 140 MeV of alpha particle and 0.002 to 210 MeV n^{-1}.

Low-energy particle experiment (LEP): Magnetotail dynamics are studied to infer plasma characteristics, its circulation, variability, response to fluctuations in the solar wind and interplanetary magnetic field of range of 0.007–42 keV Q^{-1}, the 3-D velocity distributions of solar electrons in the energy range of 6–36 eV, positive ions in the magnetic field. The sensors are assembled to measure the 3-D velocity distributions of hot plasma in the magnetosphere of electron concentration in the energy range of 0.007–42 keV Q^{-1}. The 3-D velocity distributions of solar wind ions in the energy range of 0.18 keV Q^{-1} for 3-D determination of the ion composition in the energy range of 0–25 keV Q^{-1}.

Plasma waves investigation (PWI): This sensor is essential to study the wave phenomena related to plasma dynamics in the different regions on various scales. It includes magnetic field line merging, moving plasmoids, and particle acceleration. Part of the package of the instrument is used for the navigating of the magnetosphere in elliptical orbit by the magnetotail.

Energetic particle and ion composition element (EPIC): It measures the charge, mass energy, and ions to study the relative importance of ions sources and mechanism of acceleration, transportation and loss of particles, the transportation and dynamics of magnetospheric boundary layers. Assemblages of suprathermal ion composition spectrometer to measure charge state, mass, and energy of ions with energies of 30–230 keV $charge^{-1}$; and an ion composition subsystem to measure mass and energy properties of energetic ions with energies ranging 50 KeV to 3 MeV z from 1 to 8.

Comprehensive plasma investigation (CPI): This sensor also measures 3-D plasma of Earth magnetotail. The plasma data are correlated with the magnetic field, plasma waves, electric particles, and auroral electric data to determine magnetotail in the range of plasma dynamics. Hot plasma and ion composition analyzer in the range of 1 eV to 50 keV with Z ranging from 1 to 8 and solar wind analyzer in the range of 0–15 keV $particle^{-1}$.

Ring current made of charged electrons are formed by space weather fluctuation activities. The fluctuation of the ions with capacity of a few keV to several hundred keV plays an important role in regulating the magnetic storms in Sun–Earth environment, which holds a key in deciphering and early warning of various disasters on the Earth. The instability of the ring current plasma originates Whistler wave, magnetosonic, and EMIC waves which can heat or pitch angle scatter other plasma populations through wave particle interactions and results in plasma energization. It leads to fluctuation in the ionospheric conductance. It is required to launch extraterrestrial satellites to monitor ionospheric perturbation, which is a precursor of various disasters including earthquakes.

The purpose of DEMETER satellite was to infer ionospheric perturbation during coronal mass ejection and solar flare emission from the Sun. DEMETER is the most effective scientific and dedicated satellite for the detection of electromagnetic signals associated with telluric activity. Earthquake and volcanic eruption prediction, one of the biggest challenges for scientists, being done at the ground level or from space has pointed out the existence of low-intensity magnetic and electrical anomaly. It is established that increase in the signal intensity of ionic changes is recorded by the satellite sensor of DEMETER during the final step of preparation for earthquakes that is between a few tens of minutes and a few hours prior to the triggering of the earthquake. These observations needed to be verified by other satellite sensors and ground truth measurements. Best result of the earthquake prediction depends on the following observations:

i) Continuous data collection
ii) In the appropriate frequency ranges
iii) In the different components of the electromagnetic field.

Observations from the DEMETER satellite not only were used for the development of precursor algorithms but also help to systematically study the emissions of electromagnetic waves observed during earthquakes and volcano eruptions, the disturbances of the ionosphere, and atmospheric changes including pressure and precipitations of associated particles.

It is worth mentioning here that not only CME but sudden solar flares during solar minimum may also trigger natural hazards including earthquakes. On 11 February 2009, an earthquake of magnitude 6.0 on the Richter scale was recorded in La Aquila, Italy, which was preceded by solar flare and electromagnetic field anomaly observed by the DEMETER satellite (Figures 3.5 and 3.6).

Solar Cycle 24 has shown solar minimum; however, in 2020 Cycle 25 began. It is essential to monitor the solar minimum and maximum by different sensors including ionospheric monitoring devices. In a press event, experts from the panel, NASA, and NOAA discussed the analysis and Solar Cycle 25 prediction, and how the rise to the next solar maximum and subsequent upswing in space weather will impact our lives and technology on Earth. A new solar cycle comes roughly every 11 years. Over the course of each cycle, the star transitions from relatively calm to active and stormy, and then quiet again; at its peak, the Sun's magnetic poles flip. Now that the star has passed solar minimum, scientists expect the Sun will grow increasingly active in the months and years to come. Understanding the Sun's behavior is an important part of life in our solar system. The Sun's

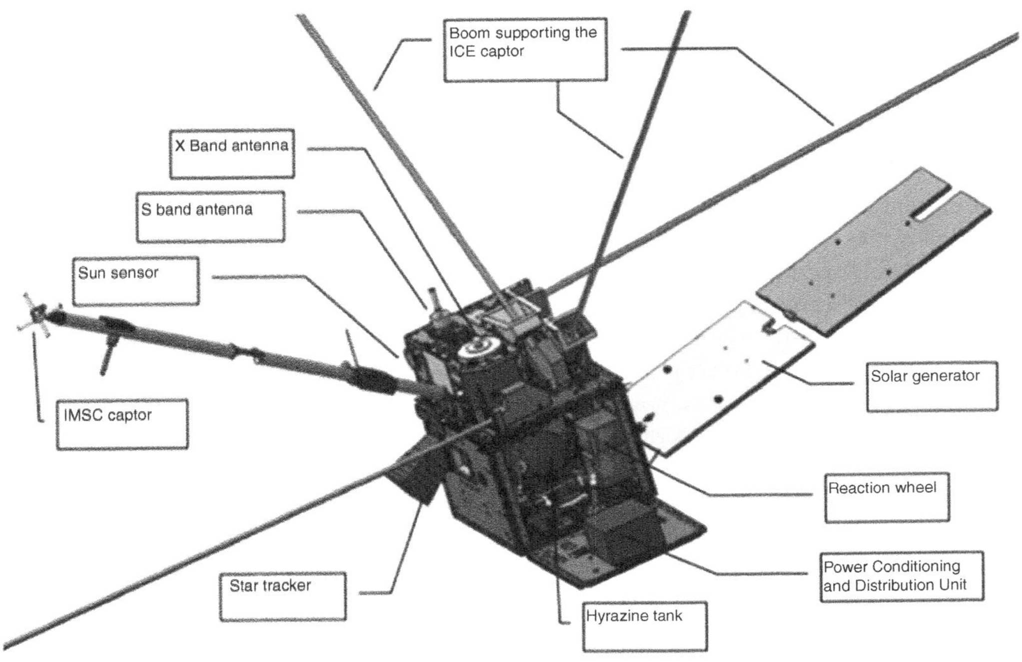

Figure 3.5 Details of the sensor system of the DEMETER satellite.

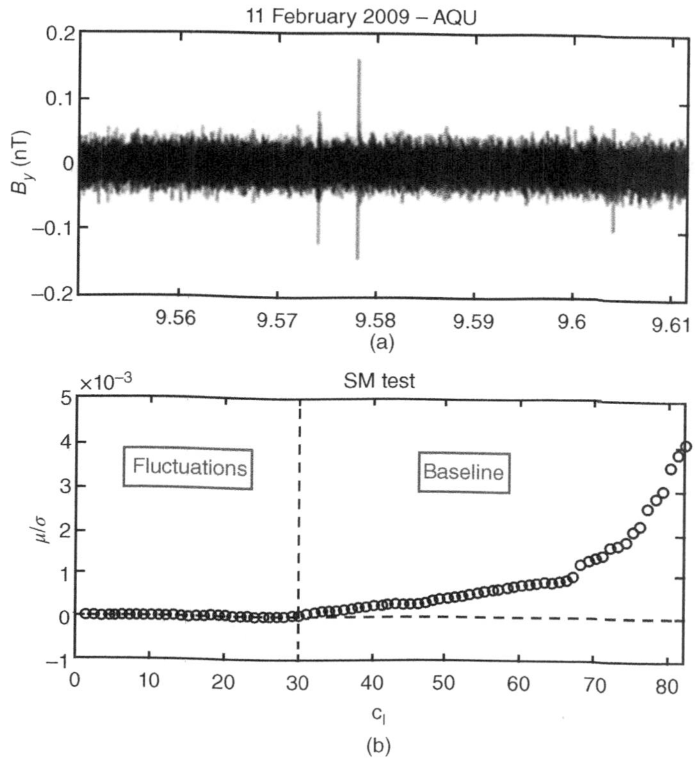

Figure 3.6 Electromagnetic field observation by DEMETER satellite showing fluctuations of safe magnetic field from its normal. Source: After Bertello et al. [1]/Public Domain CC BY 4.0.

outbursts – including eruptions known as solar flares and coronal mass ejections – can disturb the satellites and communication signals traveling around Earth, or one day, Artemis astronauts exploring distant worlds. Scientists study the solar cycle, so we can better predict solar activity. Besides monitoring the ionosphere in space weather, efforts were made to track the path of different terrestrial and extraterrestrial satellites by the Satellite Tool Kit (STK) system, which provides a measure of vulnerability (Vanguard Number) of possible collision. Space Object Threat Assessment (SOTA) helps you determine the risk of a threat object targeting another object. It provides a measure of vulnerability (called the Vanguard Number) that characterizes how likely a specific action is and how soon it could happen.

Main tasks of the analysis are as follows:

- Analyze potential activity.
- Support missions in real time.
- Fuse simulated and live data for training and exercises.
- Assess constellation vulnerability.
- Identify the most likely targets of rendezvous or intercept activity.
- War gaming and operational exercises.
- Analyze alternatives when planning threat avoidance maneuvers.

The SOTA product uses automated vulnerability analysis to ensure mission assurance for space-based systems. It is designed to support real-time, live, or simulated data, or offline for

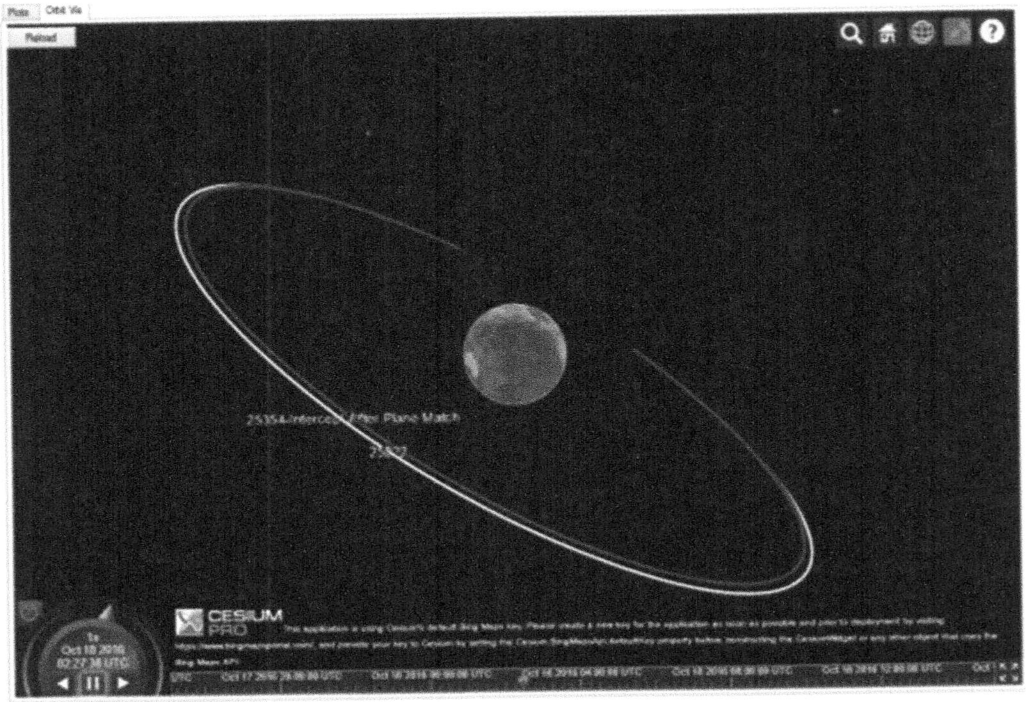

Figure 3.7 The space object threat assessment (SOTA) product.

analytical studies, training, and exercise activities. STK SOTA provides a software engine with time-dynamic web-based displays that include color-coded visualization and output for ease of sharing SOTA results. It is essential to increase the potential of STK and SOTA for safe operation of terrestrial and extraterrestrial satellites (Figure 3.7).

Reference

1 Bertello, I., Piersanti, M., Candidi, M. et al. (2018). Electromagnetic field observations by the DEMETER satellite in connection with the 2009 L'Aquila earthquake. *Ann. Geophys.* 36: 1483–1493, https://doi.org/10.5194/angeo-36-1483-2018.

Further Reading

Chilingarian, A., Hovsepyan, G., Arakelyan, K. et al. (2009). Space Environmental Viewing and Analysis Network (SEVAN). *Earth Moon Planet* 104: 195–210. https://doi.org/10.1007/s11038-008-9288-1.

Domingo, V., Fleck, B., and Poland, A.I. (1995). The SOHO Mission: An Overview. In: *The SOHO Mission* (ed. B. Fleck, V. Domingo and A. Poland). Dordrecht: Springer. https://doi.org/10.1007/978-94-009-0191-9_1.

Khachikyan, G., Inchin, A., and Lozbin, A. (2012). Spatial distribution of Seismicity: relationships with geomagnetic Z-component in geocentric solar magnetospheric coordinate system. *Int. J. Geosci.* 3 (5): 1084–1088. https://doi.org/10.4236/ijg.2012.35109.

Mukherjee, S. (2008). Cosmic influence on Sun-Earth environment. *Sensors* 2008 (8): *7736, 7752.* https://doi.org/10.3390/s8127736www.mdpi.com/journal/sensors.

Zhima, Z., Hu, Y., Piersanti, M. et al. (2020). The Seismic electromagnetic emissions during the 2010 Mw 7.8 Northern Sumatra earthquake revealed by DEMETER Satellite. *Front. Earth Sci.* 8: 572393. https://doi.org/10.3389/feart.2020.572393.

4

Extraterrestrial Satellite Data Collection and Interpretation

4.1 Introduction

An observational and theoretical information system strengthens by plotting extraterrestrial satellite data and its correlation by overlaying the information for different terrestrial solutions. Solar variables are recorded in different planetary systems including Earth. A correlation of the environmental perturbations is possible to infer based on correlations of other planets. Experiments were developed at NASA missions such as Skylab, Spacelab ATS, STS, and ATLAS satellite-based data collection and interpretation system. These satellites are orbiting in their specific orbits and most of them are using solar panels as substitutes for hydrocarbon or other gaseous fuel. Heliophysical and other valuable information data are collected in Teledyne system as charged coupled devices that are manufactured by Bismuth–Tellurium–Germanium alloy, which transfers the scintillations to the Earth receiving and monitoring stations. These heliophysical variables are being updated continuously in 15-minute intervals. Energy particle fluxes are monitored to infer Sun–Earth system. SOHO data can be processed with complicated mathematical processes; however, the output from the instrument is merged into the SOHO telemetry stream transmitted to the Deep Space Network and then delivered to the Large Angle Spectrometric Coronagraph (LASCO) workstations at the Experiment Operations Facility. For about 12 hours a day, SOHO data are received by the receiving station in near real time, and for the remaining time, these are received with a 2-hour delay. Every day the previous day's real-time and playback raw data are uploaded in FTP from the processing system. All the datasets are designated quick look, which can be downloaded through the designated website. Various ground stations record the information onto tape and mail the tapes to the concerned laboratories. These tapes are processed by NASA observatories and further forwarded to the LASCO team several months after the info was collected. The Level Zero ("LZ") data have fewer missing data packets (blocks) than the fast Look data. Raw data are processed to separate the telemetry stream into compressed image files. These files are then decompressed, oriented in order that solar north is at the highest of the image, then written out as a FITS file. Because they have been "rectified" to place solar north at the highest of the image, this means that rows and columns may have been transposed, depending upon the readout port and the orientation of the telescope on the spacecraft. Calibration images (calibration lamp images, dark current images, or pseudo-continuous readout images) are not rectified. Note that SOHO is maintained in order that the vertical axis is perpendicular to the solar equator; however, there's a small roll angle offset of every one of the LASCO telescopes from solar north. Level 0.5 includes no corrections for instrument response, stray light, etc. and are in units of DN (digital number or counts).

Level 0.5 images may be calibrated into physical units (e.g. mean solar brightness for C1 and C2). The calibration involves corrections for the flat field response of the detector, radiometric

Extraterrestrial Remote Sensing and Climate Change, First Edition. Saumitra Mukherjee.

sensitivity, stray light, geometric distortion, and vignetting a single Level 0.5 image. While the extent 1 calibration is often applied to Level 0.5 derived from quick look data, it's generally applied only to the LZ-derived FITS files. The calibration is still not final; however, it is possible to obtain calibrated images using reduce_level_1.pro in the LASCO section of the solar soft IDL library.

Level 2 (FITS): Processing involving multiple images, like a polarization, color analysis, or on/offline combinations, are considered Level 2 processing. Such processing may involve grouping images by time, color filter, polarizer, or wavelength. The Level 2 products are produced directly from Level 0.5 FITS files for a subset of images.

Besides solar variability studies, the planetary system variability also depends on the solar changes. The influence of the Sun is often observed in other planets of the system also. The Saturn ring has provoked scientists and armature astronomers since the advancement of human civilization. Various hypotheses have been postulated on the connection of sunspots, Earth's environment, and changes in the Saturn ring. A concept was given on Planetary-Spin heat, Jupiter–Saturn–Solar Tidal–Sunspot coupling, and Planetary–Solar alternating magnetic field rings by Chris Landau on 13 September 2010. It shows how our planets and stars interact through tidal and magnetic forces, and three new concepts are proposed:

(a) Heating our Earth and the planets by rotation and spin, not by radioactive decay.
(b) Jupiter and Saturn's Spring Tides cause the Sunspot Cycle and Solar Magnetic field reversals.
(c) Planetary rings of gas giants are caused by alternating magnetic fields due to a rapid spin.

The relationships are:

1. Rapid planetary rotation causes internal heating through density increase, changing magnetic and electric flows within the core, resulting in convective heat, continental drift, and volcanism.
2. Sunspot Jupiter–Saturn Tidal link for sunspot cycle timing and solar magnetic field reversals.
3. Rapid spin of planets and stars (pulsars) leading to alternating magnetic fields, responsible for ionized planetary rings and torus rings. These are suspended halfway between the poles of planets and pulsars. Alternatively, pulsing beams of high-energy particles from pulsars are due to the alternating magnetic fields within their cores.

Tidal forces of Jupiter and Saturn may control our Sunspot Cycle. Our two gas giants Jupiter and Saturn have a large tidal influence on our Sun, much like our moon has a very large tidal influence on our oceans, our atmosphere, and our solid Earth. These tides where we have competing gravitational forces between the Sun and Moon cause our daily measureable influence on our 12-hour cycle of the tides on our oceans and large inland seas, our weather, due to compression and expansive forces and even small earthquakes and minor volcanism changes as documented on Mount Etna in Sicily. The tides on our Sun's atmosphere or photosphere by Jupiter and Saturn affect Earth's atmosphere too. As Jupiter and Saturn revolve around the Sun in 11.85 years for Jupiter and 29.5 years for Saturn, they are either on the same side of the Sun reinforcing the tidal effect (equivalent to the Spring tide on Earth–Moon–Sun system) or on opposite sides of the Sun, also known as Spring tide. This occurs every 10.7 years approximately. Because of Jupiter's mass, the Jupiter–Sun center of gravity position falls 7% outside our Sun and has a great tidal effect on the Sun. So here we have a simple answer for what causes the pull on our Sun and distorts that atmosphere to create solar dark spots that are actually magnetic field storms. These storms occur because the Sun is composed of plasma (free electrons, protons, and neutrons moving in circular currents, much like our Earth's atmospheric storms move in circular fashion, driven by heat and the Carioles force). These magnetic storms ionize our Earth's and other planets' atmosphere and probably provide more condensation nuclei in the form of charged, dust, salt, and smoke particles

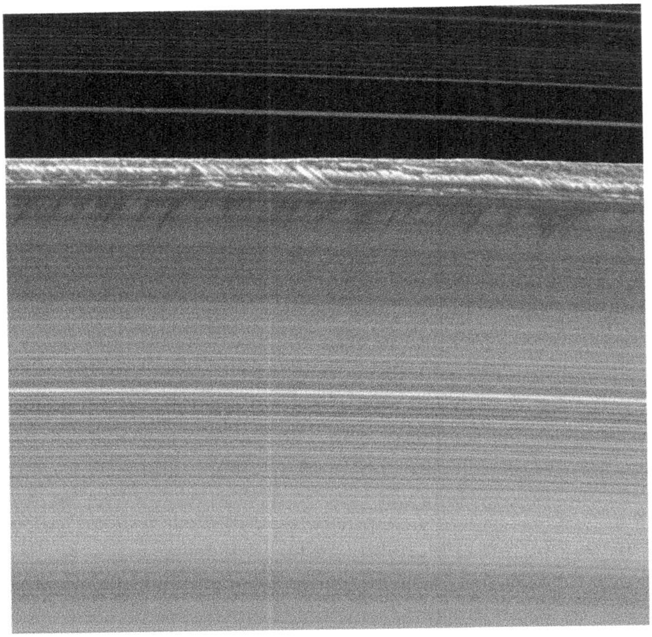

Figure 4.1 Saturn ring. Source: Courtesy: NASA/JPL/Space Science Institute.

that either increase or decrease our rainfall over the same 11-year cycle. Planets' rotational spin heats and cools our planet's core to provide the answers to some of the magnetic reversals we see. These magnetic reversals should also be read the other way around and tell us what is happening in our Sun and what is still going to happen. They provide clues to changing events. I think that once we know where all the orbiting bodies in our solar system are, we will understand the big picture that controlled all of these past recorded events, which are written in the pages of our planets. We have only just begun to read the pages in the other planets of our solar system. Saturn's rings are a magnetic plane of ionized particles suspended halfway between the two poles as the magnetic field oscillates and alternates. This plane is not a torus but a ring. Saturn's rings appear to form and disperse within only a few hours. Scientists have conjectured that these spokes might be composed of electrically charged sheets of dust-size particles created by small meteors impacting the rings or by electron beams from the planet's lightning (Figure 4.1).

Saturn observation on 30 June 2004 was remarkably noted in view of Sun–Saturn–Earth environment. There was a sharp rise in global temperature from 28 June 2004, which was found in an increasing trend till 2 July 2004. A sharp rise in E-flux, P-flux, and Kp indices were noticed 36 hours before coronal mass ejection (CME) from the sunspot. It has been observed that the Saturn ring has been showing some unusual waves in the Cassini latest images. Saturn rings are made up of a mixture of dust-ice-particles, which are influenced by the sudden changes in Sun–Saturn environment. Similar observations are being made on Earth. Thermosphere–ionosphere–atmosphere and geosphere of Earth were found to be changing during CME from the Sun. During the global temperature rise from 28 July to 2 July, the biomass of various parts of Earth were found disturbed. Large numbers of snakes, rats, and other animals living inside the Earth came restlessly to the surface of the Earth. From 2 July, the global temperature decreased as well as it is found that a wave in the Saturn ring was reduced (Figure 4.2).

Figure 4.2 Reduced Saturn ring during solar maximum. In another 100 million years, the rings may disappear altogether.

Table 4.1 Daily geomagnetic data of 28 June to 2 July 2004.

Middle latitude	High latitude	Estimated				
	- Fredericksburg -	---- College ----	--- Planetary ---			
Date	**A**	**K-indices**	**A**	**K-indices**	**A**	**K-indices**
2004 06 28	11 2 4 1 2 2 1 3 3	-1 1 1 0 1 1 0 0-1	13	3 3 1 2 2 2 4 4	High	
2004 06 29	15 3 3 3 2 2 4 3 3	29 5 5 4 4 5 3 3 2	20	4 4 4 3 3 3 4 3	High	
2004 06 30	8 3 2 3 2 2 1 2 2	21 3 2 5 4 4 2 2 4	10	3 2 3 3 3 2 3 2	Moderate	
2004 07 01	9 2 2 3 2 2 1 3 2	24 2 3 5 6 3 3 2 2	13	3 3 3 3 3 3 3 2	High	
2004 07 02	-1 -1-1-1-1-1-1-1-1	-1 -1-1-1-1-1-1-1-1 -1 2 3 2-1-1-1-1-1	Low			

This image shows three density waves in Saturn's A ring. It was taken by the narrow angle camera on the Cassini spacecraft on 1 July 2004 after successful entry into Saturn's orbit. The view shows the dark, or unlit, side of the rings. Daily regular magnetic field variation arises from current systems caused by regular solar radiation changes (Table 4.1). Other irregular current systems produce magnetic field changes caused by the interaction of the solar wind with the magnetosphere, by the magnetosphere itself, by the interactions between the magnetosphere and ionosphere, and by the ionosphere itself.

Besides Kp indices, X-ray flux was also high during this period, which affected the Saturn ring (Figure 4.3).

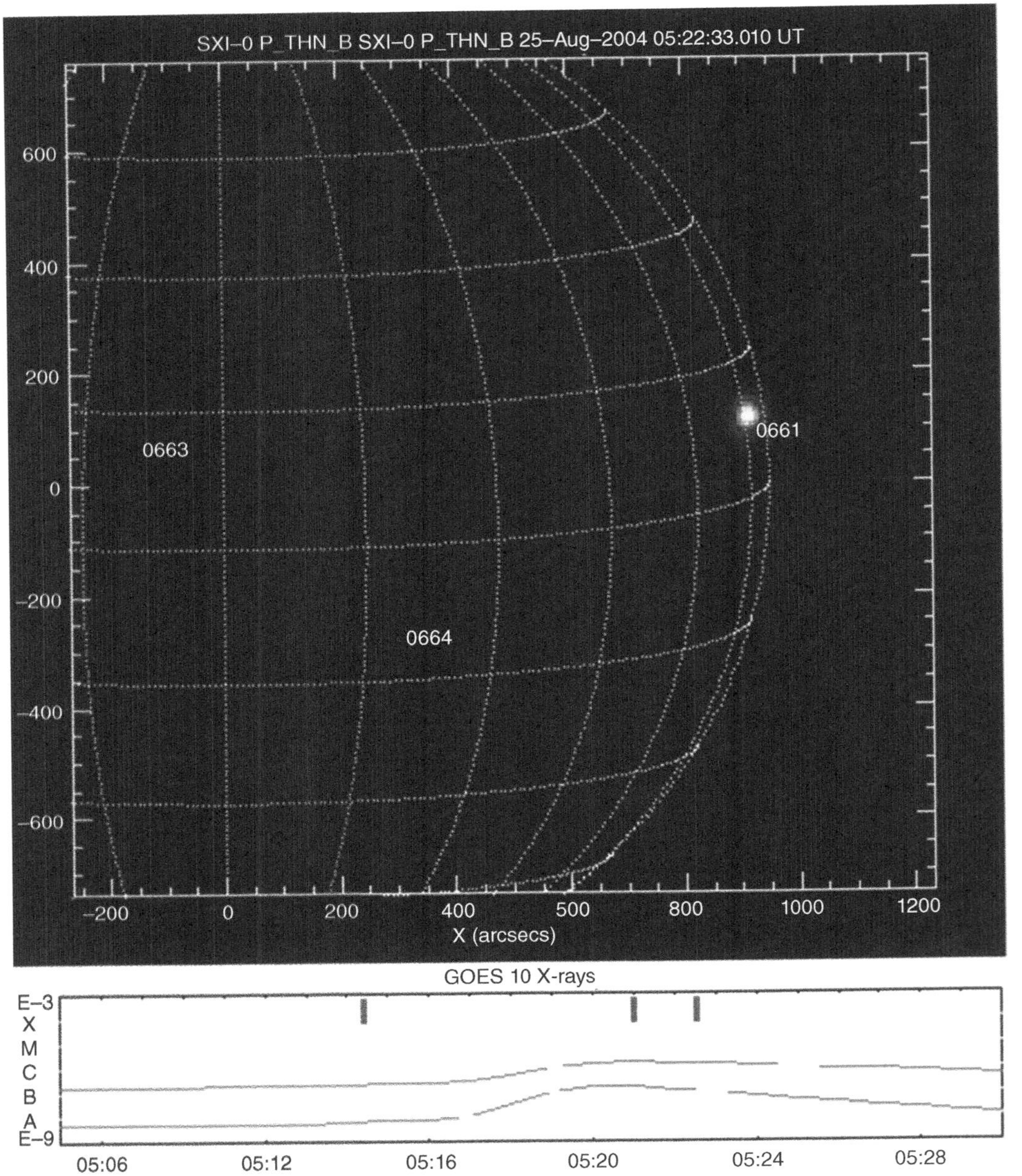

Figure 4.3 Continuous raised X-ray flux, which leads to change in the Saturn ring. Source: Adapted from NOAA https://www.swpc.noaa.gov/.

There is a possibility that the Saturn ring will disappear if solar plasma emissions are greater. During the solar prominence like CME or solar flare if it is Earth-directed, it will destroy the environment of the Earth while the Saturn rings are likely to be abolished due to solar maximum. Nature rejuvenates itself; the extended solar minimum has temporarily slowed down the process of destruction of the Saturn ring and Earth's environment. Within terrestrial Earth, we may see the

Figure 4.4 Saturn ring size increasing during solar minimum. Source: zenobillis/Adobe Stock.

change in climate while Saturn, Jupiter, and other planets are also influenced by solar minimum and maximum (Figure 4.4).

The graph below shows us the number of C-, M-, and X-class solar flares that occur for any given year. It gives us a nice idea of the number of solar flares in relation to the sunspot number. It is thus another way of seeing how a solar cycle evolved over time. This data comes from the NOAA SWPC and is updated daily (Figures 4.5 and 4.6).

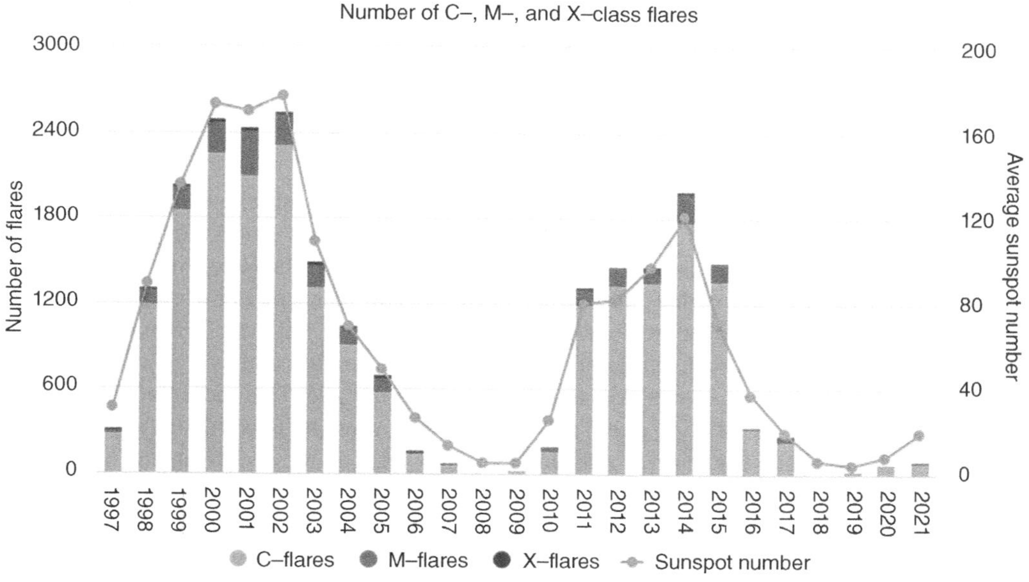

Figure 4.5 The graph shows the number of C-, M-, and X-class solar flares that were produced during July 2021 together with the sunspot number of each day. This gives an idea of the solar activity during the past month. This data comes from the NOAA SWPC and is updated daily. Source: Data from NOAA https://www.swpc.noaa.gov/.

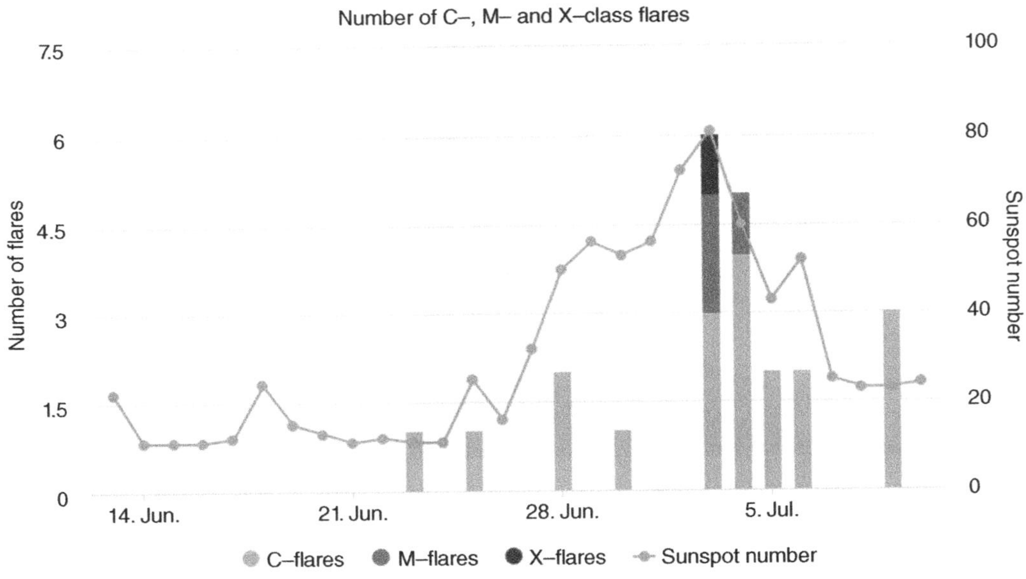

Figure 4.6 During extended solar minimum sudden rise in sunspot number in July 2021. Source: Data from NOAA https://www.swpc.noaa.gov/.

The sunspot activity influences temporarily the magnetic field of the Earth, which is recorded by SOHO satellite. However, the short-term changes of the heliophysical parameters cannot change the density of ice and dust particles of the Saturn ring.

4.2 The Sunspot Activities

4.2.1 Solar Flares Behave Like a Twinkling Butterfly

Throughout the solar cycle, the latitude of sunspot regions varies with an interesting pattern. The graph below shows the latitude of all sunspot regions of the last 22 years versus the time (in years). Sunspots are typically confined between −35° south and +35° north latitude. At the beginning of a new solar cycle, sunspot regions are formed at higher latitudes, but as the cycle progresses toward the maximum, the sunspot regions gradually form at lower latitudes. When nearing the solar minimum, the sunspot regions appear around the solar equator, and as a new cycle starts again, sunspots of the new cycle will start to emerge at high latitude. This recurrent behavior of sunspots give rise to the "Butterfly" pattern and was first discovered by Edward Maunder in 1904. Periodic changes in the sun has been recorded in 11 year as well as daily cycle. The graph is updated every month (Figure 4.7).

Solar flare and CME may have a role in total electron content (TEC) in different planetary upper layers. In all the planets, the atmosphere is not like the Earth; however, the impact on Earth can be seen as a correlation of lightning intensity and high TEC. The TEC in the ionosphere is modified by changing solar extreme ultraviolet radiation, geomagnetic storms, and the atmospheric waves

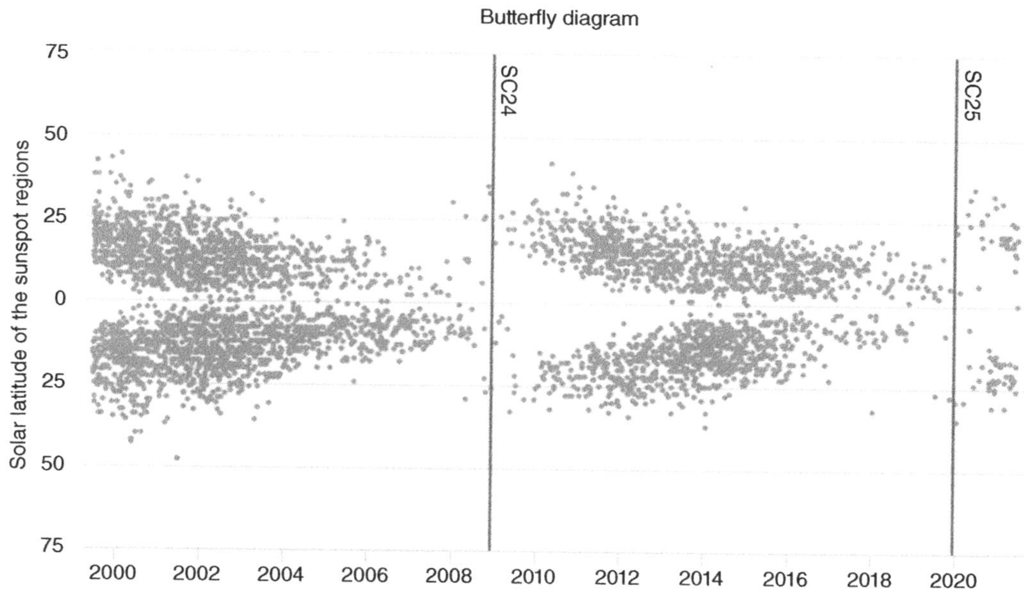

Figure 4.7 Butterfly diagram showing low sunspot numbers in 2009 and 2019–2020. Source: Adapted from P. Charbonneau and O.R. White–April 18, 1995.

that propagate from the lower atmosphere. The TEC will therefore depend on local time, latitude, longitude, season, geomagnetic conditions, solar cycle and activity, and troposphere conditions. The propagation of radio waves is affected by the ionosphere. The velocity of radio waves changes when the signal passes through the electrons in the ionosphere. The total delay suffered by a radio wave propagating through the ionosphere depends both on the frequency of the radio wave and the TEC between the transmitter and the receiver. At some frequencies, the radio waves pass through the ionosphere. At other frequencies, the waves are reflected by the ionosphere.

A case study of sudden change in heliophysical parameters on 5 July 2021 has been manifested as a rise in electron flux on 10 July 2021 and lightning in Jaipur, Rajasthan, India, on 11 July 2021 (Figure 4.8).

Dark circles are showing location of intense lightning from Jaipur to Prayagraj, India, on 11 July 2021. The plot illustrates the height-integrated electron density (TECU, 1 TECU = 1.e16 electrons/square meter) also called vertical total electron content (VTEC) vs. latitude (−90 to 90°) and longitude (0–360°) from the Coupled Thermosphere Ionosphere Plasmasphere Electrodynamics Model (CTIPe). Similar correlation has been made across the world including Parana, Brazil, and Southern Africa.

CTIPe is a state of the art research tool used at the Space Weather Prediction Center to study thermosphere–ionosphere phenomena in order to develop now casting and forecasting algorithms for space weather. The objectives are to understand and quantify the importance of the upper-atmospheric mechanisms that affect human activities and to develop new monitoring and predicting techniques. The CTIPe model runs are performed using real-time ACE measurements from the L1 point to infer the necessary inputs. Because the solar wind takes 30–40 minutes to propagate from L1 to the nose of the magnetosphere, the model results are displayed on the web

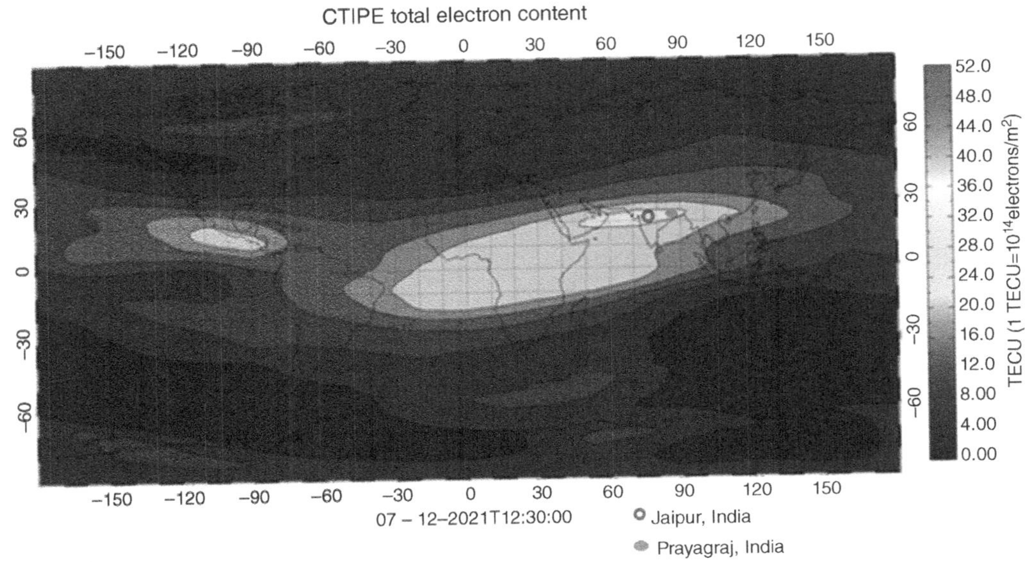

Figure 4.8 Higher total electron content formed due to changes in space weather has increased the chances of lightning from Jaipur to Prayagraj, India, on 11 July 2021. Source: Adapted from https://www.sws.bom.gov.au/Satellite/2/2.

page 20–30 minutes ahead of real time. This is an average time as we do not take into account the current solar wind speed in its calculation. The F10.7 for the previous day averaged with the F10.7 average over the previous 41 days is used to drive the heating, dissociation, and ionization processes. One-minute ACE data is used to generate the high-latitude electric potential patterns using the Weimer model and to infer the particle precipitation patterns using correlations between ACE data and the model of auroral precipitation patterns. The input values (solar wind Btot in the YZ plane, solar wind velocity, IMF angle, solar wind density, and tilt) used for the Weimer model and to infer the particle precipitation pattern are obtained from the SWPC real-time development data base (Figures 4.9 and 4.10).

On 9 July 2021, the sunspot is producing only B-class X-ray flux but on 11 July the X-ray flux was much higher and very near to C-class X-ray flux.

Rise in X-ray flux have an impact on the ionosphere to generate free electrons, which effectively raised the total electron flux in near Earth environment.

A lightning strike killed at least 16 people and injured many more from Jaipur to Prayagraj in northern India on Sunday, 11 July 2021.

The victims were taking selfies in the rain on top of a watchtower at the city's 12th Century Amer Fort, a popular tourist attraction. Dozens have also died in lightning strikes in Uttar Pradesh and Madhya Pradesh states. Lightning strikes kill some 2,000 Indians on average every year, according to official data. It is essential to develop a system to predict the space-weather-based natural disasters to minimize the human loss.

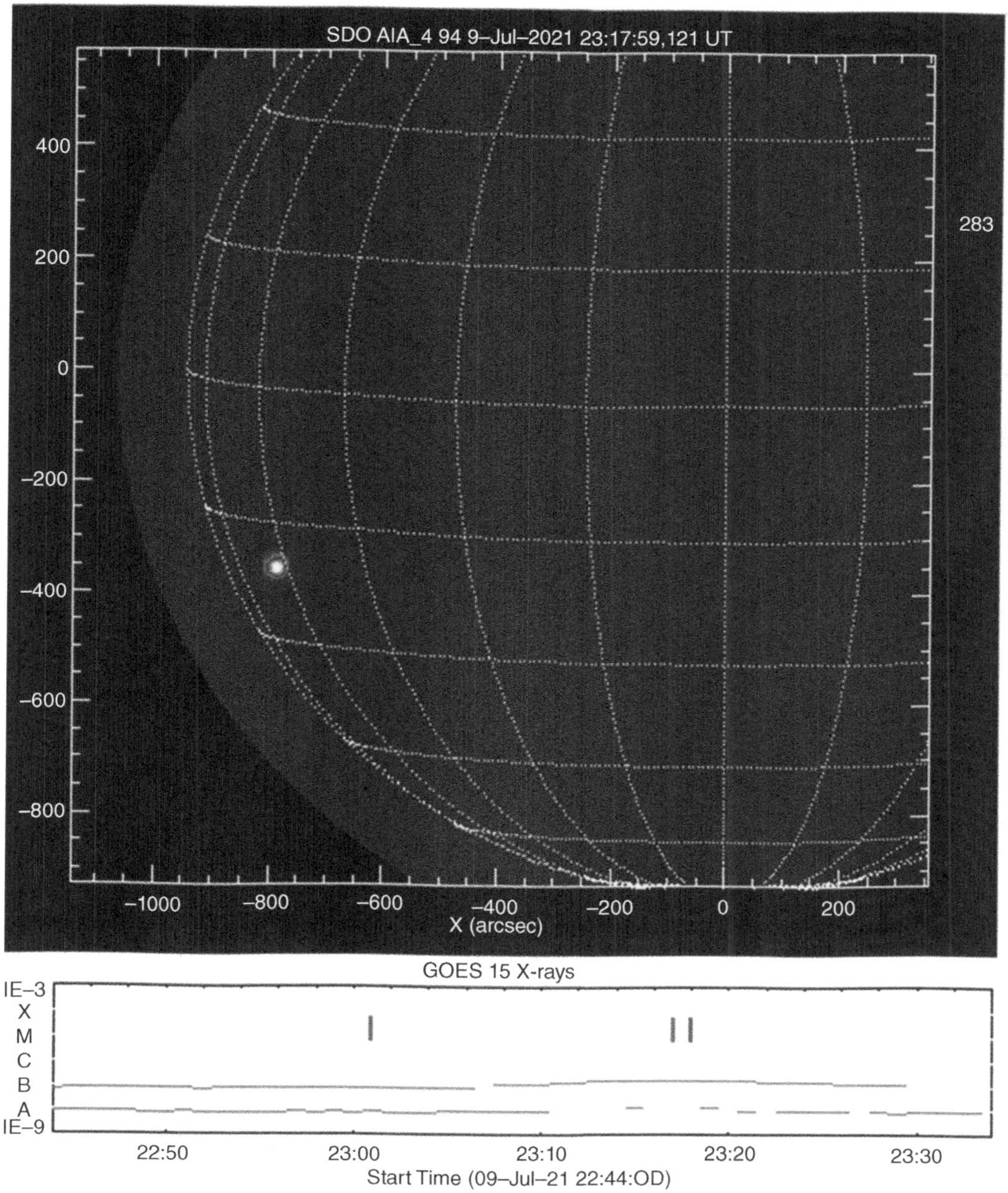

Figure 4.9 SOHO data showing rising sunspot and above threshold X-ray flux on 9 July 2021. Source: Adapted from https://www.swpc.noaa.gov/.

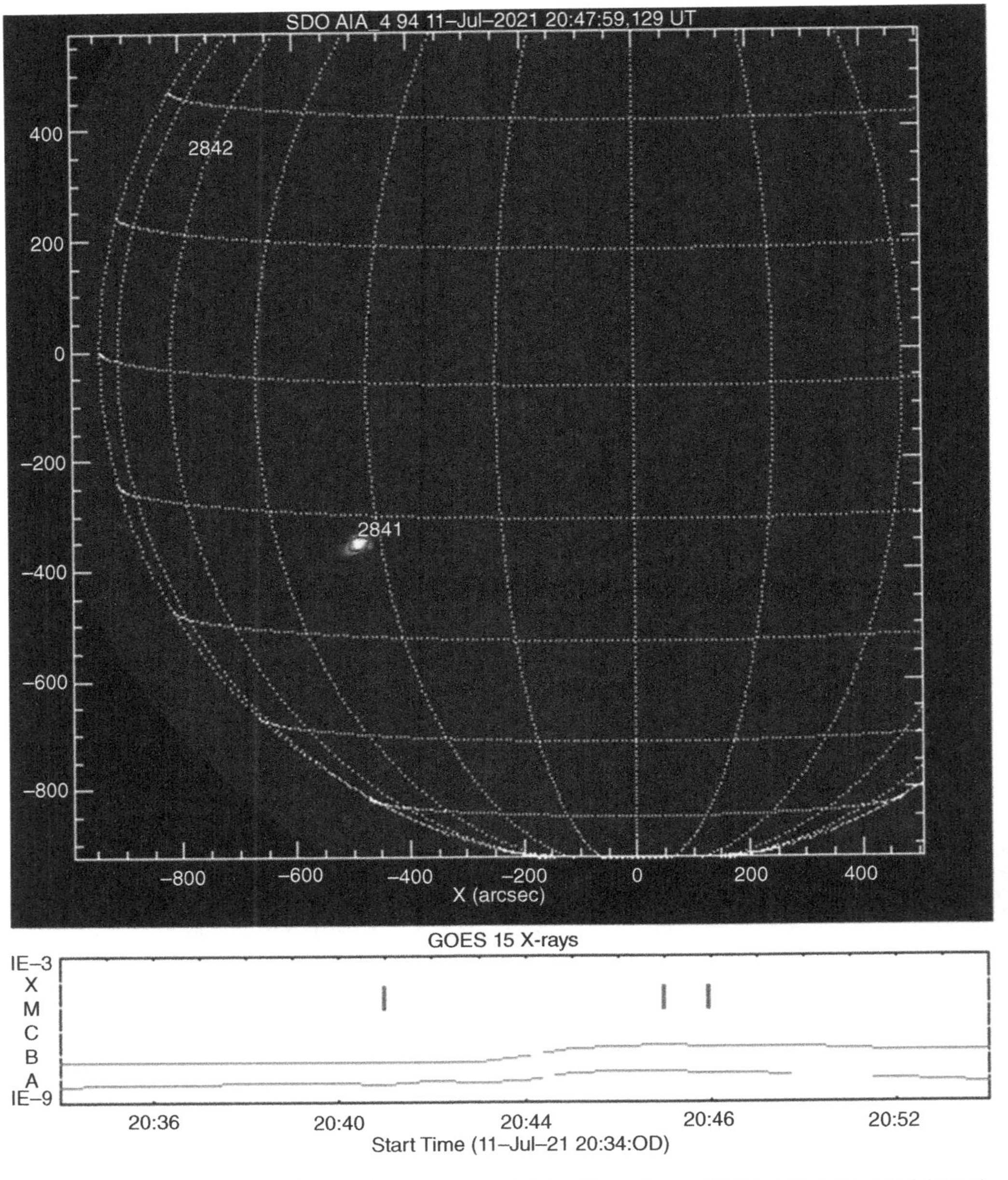

Figure 4.10 SOHO data showing moving sunspot and rising X-ray flux on 11 July 2021. Source: Adapted from https://www.swpc.noaa.gov/.

Further Reading

Davis, C.J. and Johnson, C.G. (June 2005). Lightning-induced intensification of the ionospheric sporadic E layer. *Nature* 435 (9): 799–801.

Johnson, C.G. and Davis, C.J. (2006). The location of lightning affecting the ionospheric sporadic-E layer as evidence for multiple enhancement mechanisms. *Geophys. Res. Lett.* 33 (7): ISSN 1944-8007. https://doi.org/10.1029/2005GL025294.

Mostajabi, A., Finney, D.L., Rubinstein, M. et al. (2019). Now casting lightning occurrence from commonly available meteorological parameters using machine learning techniques. *Npj. Clim. Atmos. Sci.* 2: 41. https://doi.org/10.1038/s41612-019-0098-0.

Pasko, V.P., Inan, U.S., Bell, T.F., and Taranenko, Y.N. Sprites produced by quasi-electrostatic heating and ionization in the lower ionosphere. *J. Geophys. Res. Space Phys.* A3: 4529–4561. ISSN 2156-2202. https://doi.org/10.1029/96JA03528.

Tippett, M.K. and Koshak, W.J. (2018). A baseline for the predictability of U.S. cloud-to-ground lightning. *Geophys. Res. Lett.* 45: 10,719–10,728.

5

Terrestrial Satellite Data Correlation with Extraterrestrial Satellite Data

The Earth and other planets like Mars were formed at the same time; however, Moon of the Earth has different origin. Different physical processes of the Earth, Moon, and Mars appear to be the same when its tectonic activities, volcanic activities, and structure of meteorite impacts are studied. Martian studies show similar fluvial activities in the past as well. Huygens basin of Mars was examined, which penetrates the ejecta blanket of Hellas basin, thereby exposing pre-Hellas crust, which is the oldest crust in the study area. A very unique feature of Huygens is its peak ring structure, which contains morphologies signifying past fluvio-glacial activities. We observed rampart craters of comparable size showing different lobate morphologies of ejecta blankets in the same area indicative of varying concentration of volatile presence on the surface at the time of impact. Several sinuous ridges, or eskers, also seen within the peak ring are interpreted to have formed due to melting of sub-glacial tunnels filled with deposits.

Moving forward with the primary objective of exploring natural resources on Mars, it was possible to conduct a global survey using images from the Mars Colour Camera to examine valley networks. Understanding the origin of fluvial valley networks, which are indicative of liquid water, can provide us clues to elucidate past Martian climate. Difference in the morphology of the channels around Huygens despite their proximity to the apparent source led us to explore these networks further. Drainage network extraction around the Huygens crater reveals a flat basin to its north. The difference (based on morphology and morphometry) between the channels to the northwest and south of the identified basin suggests different origins. It was found that three different kinds of desiccated polygons within the area; in addition, signatures of phyllosilicates, hydroxylated silica, and water ice were also found. Water-related surface features and distribution of hydrous minerals suggest the presence of a playa. Based on these findings, the proposed area can be a prospective landing site for future exploration missions. It is important to compare the terrestrial information with extraterrestrial findings to infer the evolution of planetary system in space and time.

Lunar impact craters are important for studying lunar surface morphology because they are the most typical morphological units of the Moon and provide the information of both current and evolutionary history of the moon. There are various types of geological information available on Earth and its spectral library, which are correlated with lunar mineralogy and morphological characteristics. In the Moon, various highlands show multiple ejections from volcanic craters that have a higher percentage of titanium dioxide. Similar correlation is found in the lava flow of the Earth. Several craters were selected covering both the regions, i.e. highland and mare region from the study area, and we carried out a detailed study of each respective crater. These are Cardanus crater and Kraft crater from the mare region and Glushko, Dalton, and Kopff craters from the highland region. Cavalerius crater falls within the transition zone between the mare and highland. Cardanus crater, with a diameter of 50 km, is located in the western region of Oceanous Procellarum centered

Extraterrestrial Remote Sensing and Climate Change, First Edition. Saumitra Mukherjee.

Figure 5.1 MCC image of Mars showing Dendritic drainage channels and valleys with complex interconnected pattern created by repeated flooding of melted ground ice after volcanic eruptions in the past. Source: Chetankjain/Wikimedia Commons.

at 13.2°N and 72.4°W. The crater consists of polygonal fractures with a deep floor, central peak, and terraces along parts of the inner wall.

Chain of Craters are connected with the Catena Kraft crater, which is located in the north of fractured craters. All the six similar craters were clubbed into one class, which are sharp-edged outer rim and crescent-shaped mare deposits (Figure 5.1). The crater consists of unique morphological features like a central mound, hummocky unit, ejecta material, fractures, and smooth floors. The central mound is occupying 6.1 kilometers on the crater floor and its width varies from 2 to 5 km. It is in the form of a mass of irregular mounds formed due to low velocity impact of clustered ejecta material [1]. The hummocky unit can be observed in the east of the crater floor, which has a higher albedo than the surrounding terrain. Ejecta materials around the crater are target materials transported in the form of radial flow and are found deposited in all directions. They are proximal in nature, spread around the crater in the form of a continuous blanket. The smooth floor units are distributed in the north of Cardanus crater, which shows lower albedo and absence of mounds. The fractures present in the crater are polygonal in structure and cover the crater floor in all directions, the branching in the fracture is observed, and this bifurcated fracture extends further toward the south direction on the crater floor. Interpretation highlights tiny rock outcrops present at the base of the rim in the west direction of the crater floor, which might be due to the weathering process in the inner rim.

The fractures present within the crater floor are the surface manifestation of a class of shallow crustal intrusions in which magma-filled dikes rise vertically toward the surface from the mantle region. It laterally forms a sill or laccoliths that create overburden pressure on the periphery of the crater rim. The crater wall and rim deposits restrict the lateral growth of the sill, which forces it to grow vertically and form piston-like structures, uplifting the periphery of the crater floor. This peripheral upliftment can lead to the subsidence of the central region of the crater. The protons keep bombarding the surface of the moon due to the absence of atmosphere there. These protons get captured in the fractures, which might result in hydroxyl deposition in the subsided region. Further, the continuous bombardment of protons and subsequent sputtering process may lead to accumulation of these hydroxyl minerals, which may transform the subsided region into a reservoir for hydroxyl minerals.

5.1 Morphological Units

The crater has diverse morphological features, i.e. mound deposition, rampart feature showing distinctive fluidized ejecta material, smooth crater floor, presence of fractures, presence of crater inside the crater, and hummocky unit. Kraft crater has ejecta material deposited all around the crater rim with maximum width toward the east and northeast directions. A smooth floor can be seen within the crater, which is devoid of megaclast and hummocky units. The region shows the lowest albedo condition. The crater has the presence of hummocky units toward the east, which comprises mounds scattered across the unit. They can be easily interpreted on the crater floor due to their relatively high relief and optically brighter appearance in the NAC data.

The crater has fractures on the floor running from the north toward the south. It is interrupted by the presence of the bowl-shaped satellite crater Kraft C (diameter 12.45 km) having a sharp rim, presence of boulders on the floor, and a smooth inner crater rim. Kraft crater has multiple fracture zones on the floor that might be due to satellite crater formation. The rayed crater might have formed due to ejection of fragmented material from secondary crater impact. The rays formed are considered to be compositional rays that are due to albedo contrast between highland-rich ejecta and the underlying dark mantle floor.

Landslide features are seen in different craters formed by the slumping of crater wall materials along preexisting concentric faults that are parallel to the interior wall. This might be due to the dislodging of rock fragments as a result of either small impacts nearby or seismic activity.

Both the fractures make a half loop at the origin and then turn east and west, respectively, making a v-shaped structure.

Chandrayaan-1 m3 data was further used to understand the depth of certain minerals like olivine and pyroxene. In the case of the Cardanus crater, m3 label 1B dataset was used and further converted into reflectance dataset. The crater has abundant deposits of pyroxene indicated in blue color. The fractured floor and outer rim of the crater highlights green color showing depositions of olivine and clinopyroxene. Kraft crater lies close to Cardanus crater in the mare region of the area. The crater shows stronger band tilt value, which indicates the abundance of olivine and clinopyroxene marked in green color. The crater rim in the west shows deposition of orthopyroxene indicated in blue color.

5.2 Terrestrial Correlation

Ramgarh crater of India is a complex crater with a diameter of ≤4 km and is situated in the Vindhyan supergroup of sedimentary rocks of Meso- to Neoproterozoic age. The crater consists of an outer rim, inner rim, crater floor, and central peak. Each region has unique geological features. Gugugarh Shale, Samaria Shale, Reva Sandstone, Lower Bhander Sandstone, and Lower Bhander Limestone are the Vindhyan supergroups of sedimentary rocks covering the crater region. Five principle rock formation strata were inferred, which belong to Neoproterozoic geological age. Various geological structures like the presence of faults, joints, and lineaments have been found distributed in all directions. The Bhander and Rewa group of the Vindhyan supergroup is exposed in the crater region. The outer rim of the crater has depositions of Lower Bhander Sandstone.

The inner rim and crater floor constitute Samaria Shale and Gunurgarh Shale, respectively. The Samarian Shale has fine lamination and presence of intercalary bands of cherts, limestone, and ferruginous sandstones. The Gunurgarh Shale is found all around the depression region of the crater with calcite veins present in the structures near the entrance of the crater. The domal structure in the central region has Upper Rewa Sandstone stratum with quartz cemented sandstone showing intense deformation and fractures. Extensive fracturing and shattering were found in the domal structure.

The Lower Bhander Limestone is found distributed all along the periphery of the crater floor. It is finely laminated, arenaceous, jointed, and has shatter cone features that are indicative of impact events. The Lower Bhander Sandstone is seen around the rim of the crater, which is a massive bed of sandstone having ferruginous laminations. Intense joints, presence of fractures, and fault lines are seen in the region.

A total of 25 faults were found in the study area. The major fault runs in a NE-SW direction through the central region. It has dextral features due to which the Lower Bhander Limestone is shifted toward the right in the northeast region of the crater rim. This fault is 400 m long and crosses the central region of the structure bending 10°–20° toward the east. The second major fault lies in a NW-SE direction, which is considered to be a reverse fault that causes intra-formational repetition of Lower Bhander Sandstone. Various joints are seen in the Lower Bhander Sandstone, which is in the form of multiple-joint-striated surfaces in quartzite.

5.3 Geomorphology of Terrestrial Analogue

The crater consists of numbers of geomorphic features like buried pediment, residual hills, presence of water body, vegetation, and structural hills. The crater rim shows buried pediment with gentle sloping and its surface is covered with detritus of sandstones and a thin layer of soil. Residual hills can be seen around the outer rim of the crater. These are isolated hills having smooth surface rising abruptly and are bare with relatively no vegetation. Structural hills of Kaimur group can be seen around the inner crater rim having moderate to high elevation and covered with natural vegetation. Structural hills with Bhander group are seen outside the periphery of the crater. Valley hills of Bhander group lie in the entrance region of the crater that is formed due to unconsolidated deposition of sediments. Ramgarh pond can be seen in the central region of the crater, which covers an area of around 40.21 hectares. Kul River, a tributary of Parvati River, is seen flowing in the northeastern region of Ramgarh crater. Correlation of terrestrial analogue with extraterrestrial satellite data has been attempted to establish the origin of different living and nonliving cyclic processes of the Earth and possibly an extraterrestrial code of this unsolved puzzle (Figure 5.2).

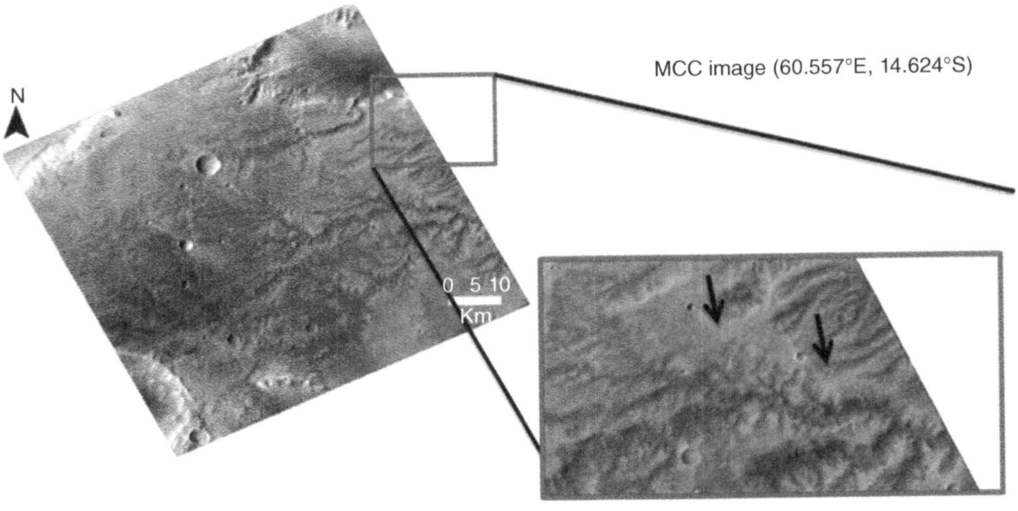

Figure 5.2 Ramgarh crater and its fluvial geomorphology. Source: Chetankjain/Wikimedia Commons.

5.4 Landform Features of Terrestrial Analogue

The topographic feature of any area depends on the elevation to a major extent. The elevation feature of the region determines various environmental aspects like weathering and erosional process, vegetation patterns, etc. In this study, the elevation map was classified based on contour lines highlighting the elevation of the study area. The maps were generated from ASTER Global DEM data with 30 m resolution.

5.5 Terrain Roughness Index

Terrain roughness index measures the terrain heterogeneities, which is based on the measurement of undulations and variations of the slope. It calculates the elevation difference between a cell and eight adjacent or neighboring cells. The slope angle of an area determines the rate of erosion. The slope map is the raster map in which each pixel denotes the angle of slope of the particular location. The slope with an angle less than 30° is considered to be slightly sloping. Maximum area is inclined toward the south and northwest directions. The aspect was derived in all directions, viz. southeast, south, southwest, west, northwest, north, northeast, and east. The curvature of the slope is formed when there is an interaction of vertical parallel plane with that of the slope direction. It controls the erosional and depositional features by influencing the flow velocities, thereby deciding the acceleration and deceleration of the flow. The positive values showed the surface to be concave and flow is accelerated. The negative value indicated the surface to be upwardly concave and flow is decelerated. The zero value indicated the surface to have linear features.

5.6 Mineralogy of Soil Samples

The mineralogy of soil samples were studied using X-ray diffraction technique. Quartz is the most dominating mineral found in all the samples. It is abundant in igneous, sedimentary, and metamorphic rocks. It is in the form of a high-pressure polymorph of quartz. It can be found in the form of isolated grains in sandstone, shale, and minor limestones. The other common minerals found in all the soil samples included orthoclase, aragonite, sanidine, sillimanite, sylvite, biotite, and wollastonite. The sample collected from the outer rim had a dominance of quartz and orthoclase mineral. The other minerals included were zircon, trona, tremolite, sanidine, spinel, carnallite, orthoclase, and muscovite.

Zircon belongs to neso-silicate minerals. Tremolites are members of the amphibole group of silicate minerals that are formed by the metamorphism of sediments rich in quartz.

Trona, or trisodium hydrogen dicarbonate dihydrate, is formed by auto-metasomatic reaction of late magma melts or by vapor unmixing in the last stages of magmatism. It is a nonmarine evaporite mineral that takes up monoclinic crystal system. It is found along with halite, thermonatrite, etc.

Muscovite, which is the most common form of mica, is basic potassium aluminium silicate. It is found in all rock types, especially in granitic pegmatites. It is formed by feldspar alteration and its presence suggests silicate weathering in the area.

Orthoclase forms due to weathering of igneous rocks. It is a tetrosilicate mineral that crystallizes in monoclinic form. Its presence is indicative of silicate weathering. It is polymorphous with the mineral microcline. Microcline is formed when monoclinic orthoclase transforms to triclinic microcline. Both of them are a kind of potassium feldspar.

The soil collected at the entrance contained primarily quartz minerals with a minor presence of aragonite, sillimanite, wollastonite, and biotite. Aragonites are polymorphs of calcite having alternate layers of carbonate groups and cations like lead, strontium, and barium. Sillimanites are aluminium silicates occurring in high-temperature regionally metamorphosed clay-rich rocks.

The central peak region of Ramgarh crater had varieties of minerals in the soil samples collected. Quartz was the dominant mineral seen in the samples. Other unique minerals found are glauconitic, rutile, tourmaline, hypersthenes, etc. Glauconitic minerals are iron potassium phyllosilicate mineral having association with sandstone and carbonate rocks. They are unstable with reference to weathering. It oxidizes to form ochre in sandstones and recrystallize into less potassic (micaceous) mixed-layer minerals in carbonate rock. Presence of glauconitic emphasizes the possibility of marine sediment in the post-crater depositional environment.

The samples collected from the soil had unique minerals like zircon, sphene, celestite, chromite, and magnetite. Magnetite is widely distributed with octahedral or spherical morphology in igneous and metamorphic rocks. Sphene is calcium titanium nesosilicate minerals occurring in igneous and metamorphic rocks. Celestite is a strontium sulphate mineral, which occurs in sedimentary rocks like limestones. Both the remote sensing and laboratory approach were carried out for in-depth investigation in the study area.

Several Earth-based studies have shown that organic matter is concentrated in anoxic bottom sets of deltaic/alluvial fans where reducing conditions exist, which promote the preservation of organic compounds from oxidative diagenesis. Hence, we explored in detail the astrobiological potential of the deltaic deposits chosen as one of the target sites for the mission. At first, we studied mineralogy of the fan to identify the best candidates for preserving the biosignatures. Mg-rich carbonate found within the upper regions of the delta is detrital in nature, implying that the carbonate deposits are of coarser grain size. Thus, carbonates present within the fan are more susceptible to radiation due to large pore spaces, which in turn makes their potential to preserve biosignatures feeble. On the other hand, since Fe/Mg Smectites have settled out at last, they are present as fine sediments and smaller pore space. Subsequently, we studied different kinds of biosignatures that can be explored in this site. The lack of oxygen in the deltaic bottom sets demands an alternate electron acceptor for the microbes to carry out their metabolic activities. Many organic/inorganic electron acceptors have been identified, but Martian crust has about twice as much iron as the Earth and clay minerals account for about half the iron in soils and sediments, which makes it a more convenient choice for microorganisms. Thus, we could expect microfossils of anaerobic chemolithotrophs, which use Fe (III) (present in smectites) as the source to generate energy, buried under the sediments. The bio-fossils may be in the form of mineralized EPS/biofilms, which are comprised of organic molecules (detectable by SHERLOC), secreted by these bacteria. Detection of elements such as Ca, Ba, Mn, and Cu embedded in the biofilm-like remnant in addition to CHNOPS by PIXL may be complementing the findings by SHERLOC.

Additionally, microbial reaction for dissolution of smectites produces illite and silica; hence, it is feasible that these deposits are also present while drilling the deltaic deposits for sample collection. Hence, we require a combination of experiments and observations to maximize the efficiency of the sample return and conclusively prove biogenicity of samples in order to bring some clarity to whether life ever existed on Mars. We started with examining Huygens basin, which penetrates the ejecta blanket of Hellas basin, thereby exposing pre-Hellas crust, which is the oldest crust in the study area. A very unique feature of Huygens is its peak ring structure that contains morphologies signifying past fluvio-glacial activities. We observed rampart craters of comparable size showing different lobate morphologies of ejecta blankets in the same area indicative of varying concentration of volatile present on the surface at the time of impact. Several sinuous ridges, or eskers, also seen within the peak ring interpreted to have formed due to melting of sub-glacial tunnels filled with deposits. Moving forward with the primary objective of exploring natural resources on Mars, we conducted a global survey using images from the Mars Color Camera to examine valley networks. Understanding the origin of fluvial valley networks, which are indicative of liquid water, can provide us clues to elucidate past Martian climate. Difference in the morphology of the channels around Huygens despite their proximity to the apparent source led us into exploring these networks further. Drainage network extraction around the Huygens crater reveals a flat basin to its north. The difference (based on morphology and morphometry) between the channels toward northwest and south of the identified basin suggest different origins. There are three different kinds of desiccated polygons within the area along with signatures of phyllosilicates, hydroxylated silica, and water ice.

Water-related surface features and distribution of hydrous minerals suggest the presence of playa. Based on these findings, the proposed area can be a prospective landing site for future.

5.7 Exploration Missions

Terrestrial analogue of Martian rocks, endorheic playa, and volcanic craters were attempted to correlate with the Dras mountain range volcanic crater and its associated land evolution

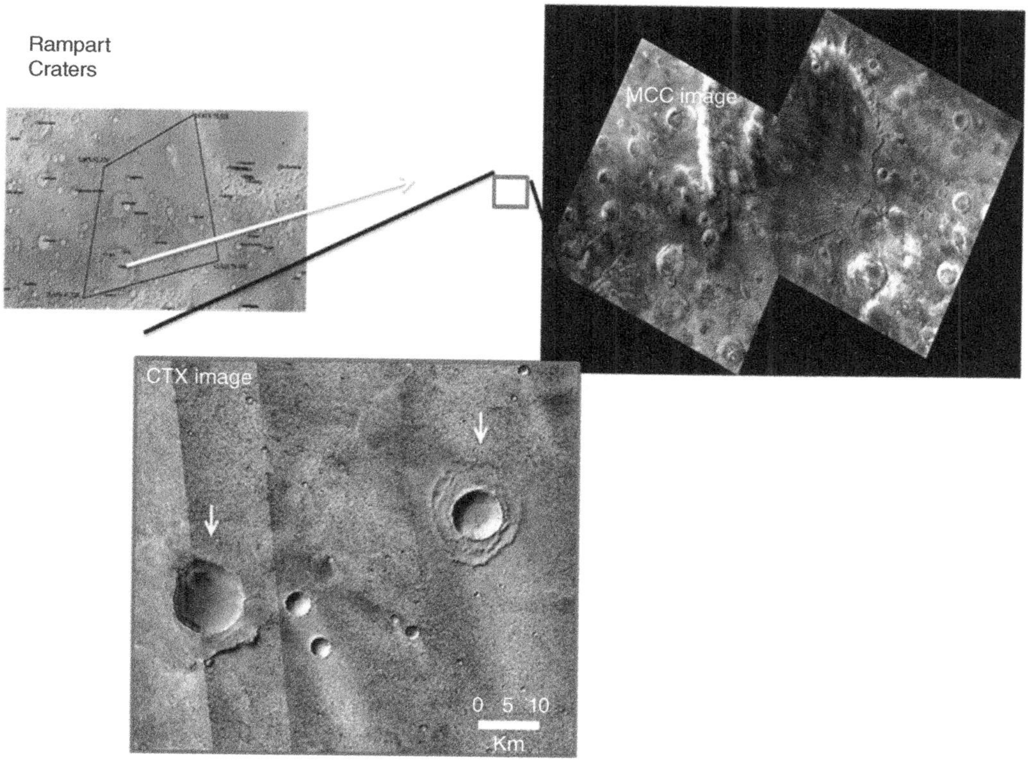

Figure 5.3 Craters in Mars inferred by CTX image.

characteristics. It is well known that the Dras mountain range is older than the mighty Himalayas, and the oldest fault of the world known as Altyn-tagh fault extends from Dras range penetrating the Indian subcontinent plunging to the Arabian Sea.

The India China border of Ladakh which is prominently known as "Indus suture zone" shows characteristics similar to the Martian Huygens crater. The tectonic activities and mineralogical evidence of the presence of glauconitic minerals in both places signify an old sea. The Eurasian Ladakh Batholithic formation and Lamayuru Complex and its arc continent collision are similar to the Martian structural anomaly near Huygens playa. The presence of felsic, andesitic volcanism early in the evolution of the Dras Arc is consistent with the adolescent stages of island arc systems, in which dehydration melting of under-plated arc or sub-ducted oceanic crust generates small volumes of felsic magmas. There are large numbers of saline lakes including Pegong Lake of Leh that are endorheic saline, which shows an origin similar to the saline lake of Mars. Both the saline lakes are formed due to tectonic activity. During winter the lake freezes completely, despite being saline water. It has a land-locked basin separated from the Indus River basin by a small elevated ridge, but is believed to have been part of the latter in prehistoric times (Figures 5.3 and 5.4).

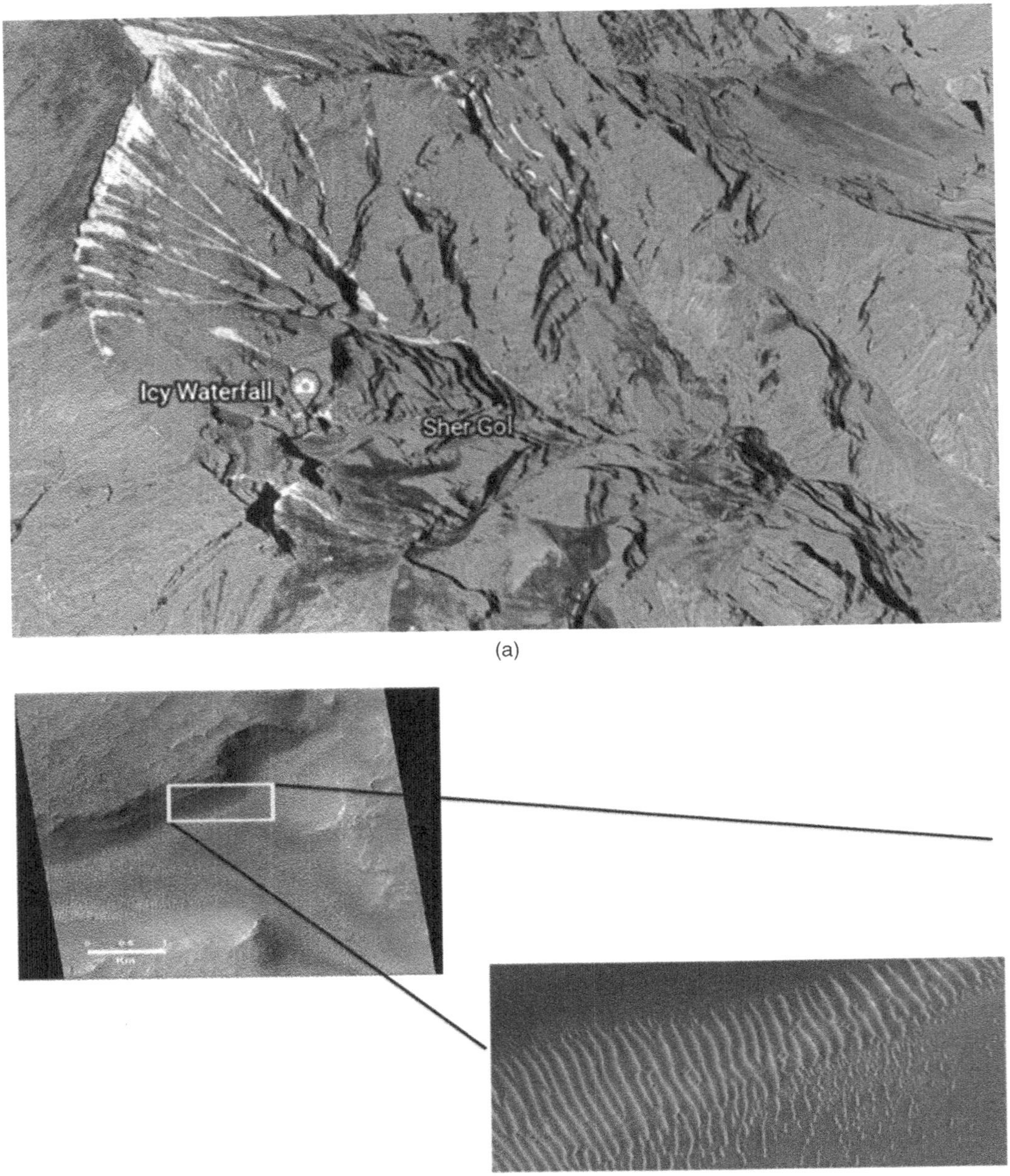

Figure 5.4 Fluvial structural regions identified (rilles, lakes beds, gullies, etc.) and their orientation to predict whether water/ice deposits are permanent features or short lived (occurring only after volcanic activity or meteorite impact). **Fluvial ripple marks**: indication of past river flows, possible presence of subsurface water. (a) Terrestrial analogue of Martian rocks, volcanic crater of Sher Gol mafic intrusive of Dras Mountain range of Leh, India. (b) Ripple Marks on Martian surface showing fluvial action.

Reference

1 Thaker, A.D., Patel, S.M., and Solanki, P.M. (2020). Morphological analysis and mapping of complex craters of Copernican age: Crookes, Ohm and Stevinus. *Planetary and Space Science* 184: 104856.

Further Reading

Borst, A., Foing, B., Davies, G., and Van Westrenen, W. (2012). Surface mineralogy and stratigraphy of the lunar South Pole-Aitken basin determined from Clementine UV/VIS and NIR data. *Planetary and Space Science* 68 (1): 76–85.

Giguere, T., Hawke, B., Peterson, C. et al. (2015). Impact melts at Glushko Crater—LROC revelations. *LPI* 1832: 1308.

Günther, A., Reichenbach, P., Malet, J.-P. et al. (2013). Tier-based approaches for landslide susceptibility assessment in Europe. *Landslides* 10 (5): 529–546.

Kenkmann, T., Wulf, G., and Agarwal, A. (2020). Ramgarh, Rajasthan, India: a 10 km diameter complex impact structure. *Meteoritics & Planetary Science* 55 (4): 936–961.

Mackenzie, F.T. (2005). *Sediments, Diagenesis, and Sedimentary Rocks: Treatise on Geochemistry, Volume 7*, vol. 7. Elsevier.

Mahanti, P., Robinson, M., and Thompson, T. (2015). Characterization of lunar crater wall slumping from Chebyshev approximation of lunar crater shapes. *LPICo* 1863: 2081.

Misra, S., Srivastava, P.K., and Arif, M. (2019). Remote sensing, structural and rock magnetic analyses of the Ramgarh structure of SE Rajasthan, Central India: further clues to its impact origin and time of genesis. In: *Tectonics and Structural Geology: Indian Context*, 327–352. Springer.

Mukherjee, S., Singh, P., Singh, D., and Roy, N. (2018a). Evidence of Glaciation based on Peak Ring morphology of Huygens Basin. In: *49th Lunar and Planetary Science Conference 2018*, LPSC 2018, No. 1925, Houston, Texas, USA. http://adsabs.harvard.edu/abs/2018LPI....49.1925S

Mukherjee, S., Singh, P., Singh, D., and Roy, N. (2018b). Landing site selection on Mars in Iapygia Quadrangle. In: *2nd International Mars Sample Return Conference 2018*, Berlin, Germany (25–27 April 2018). LPI Contribution No. 2071) https://atpi.eventsair.com/QuickEventWebsitePortal/2nd-international-conference-on-mars-sample-return/home.

Mukherjee, S., Singh, D., Singh, P., and Roy, N. (2020). Morphological and Morphometric analysis of a topographic depression near Huygens basin, Mars: identification of a putative endorheicplaya. *Geomorphology* 351: 106912.

Neish, C., Blewett, D., Harmon, J. et al. (2013). A comparison of rayed craters on the Moon and Mercury. *Journal of Geophysical Research: Planets* 118 (10): 2247–2261.

Pakhomov, E.A. (2001). Encyclopedia of Ocean Sciences – JH Steele, SA Thorpe and KK Turekian (eds). *African Journal of Aquatic Science* 26 (2): 149–149. https://doi.org/10.2989/16085910109503736.

Pareta, K. and Pareta, U. (2016). Landform classification and geomorphological mapping of Ramgarh structure, Rajasthan (India) through remote sensing and geographic information system (GIS). *Journal of Hydrology and Environmental Research* 4: 1–17.

Roy, N., Singh, P., Rena, V., and Mukherjee, S. (2021). Morphological and mineralogical characterization of Cavalerius crater in the equatorial transition region of Moon. *Ecology, Environment and Conservation* (3).

Singh, D., Mukherjee, S., Singh, P., and Roy, N. (2018c). Identification of polygonal desiccated polygons in the equatorial region of Mars: hints of recent liquid water activities. *AGU Fall Meeting 2018*,

(10–14 December 2018). EP23F-2386. https://agu.confex.com/agu/fm18/meetingapp.cgi/Paper/464959.

Srivastava, N. and Varatharajan, I. (2016). Geomorphology of Lowell crater region on the Moon. *Icarus* 266: 44–56.

Stopar, J., Hawke, B., Robinson, M. et al. (2012). Distribution, occurrence, and degradation of impact melt associated with small Lunar craters. *LPI* 1659: 1645.

Walsh, J.M.J., Buckman, S., Nutman, A.P., and Zhou, R. (2021). The significance of Upper Jurassic felsic volcanic rocks within the incipient, intraoceanic Dras Arc, Ladakh, NW Himalaya. *Gondwana Research* 90: 199–219. ISSN 1342-937X.

Whitten, J., Head, J.W., Staid, M. et al. (2011). Lunar mare deposits associated with the Orientale impact basin: new insights into mineralogy, history, mode of emplacement, and relation to Orientale Basin evolution from Moon Mineralogy Mapper (M3) data from Chandrayaan-1. *Journal of Geophysical Research: Planets* 116 (E6): 1–33.

Xiao, Z., Zeng, Z., Li, Z. et al. (2014). Cooling fractures in impact melt deposits on the Moon and Mercury: implications for cooling solely by thermal radiation. *Journal of Geophysical Research: Planets* 119 (7): 1496–1515.

6

Solar Eclipse and Climate Change

6.1 Introduction

Sun is the prime mover of different environmental cycles of the Earth including air quality. It is well established that there are changes in the atmosphere of the Earth and other planets during a solar eclipse. It is essential to monitor the impact of the Sun during solar eclipse. The effects of eclipse were observed in a part of China on 23 July 2009, which shows that the cosmic ray intensity decreases with the decrease in electron flux recorded by the Sun Heliospheric Observatory (SOHO) satellite. The attempt was to monitor the changes in the concentration of atmospheric gases like sulfur dioxide (SO_2), nitrogen dioxide (NO_2), aerosol, and cloud cover. The concept shows direct correlation of cosmic ray intensity and heliophysical and atmospheric variation during the solar eclipse. It will be useful to understand the atmosphere in variable conditions. An annular solar eclipse occurred on 10 June 2021, when the Moon passed between the Earth and the Sun. During the eclipse, there were drastic fluctuations in the atmosphere of the Earth, including a sudden loss of extreme ultraviolet radiation from the Sun, which generates the ionized layer of Earth's atmosphere, called the ionosphere (Figure 6.1).

It is our natural curiosity to look at the sky and try to understand the environment of the Earth and the Sun. It has been attempted to estimate the impact of solar variability on climate through the study of the total solar irradiance and galactic cosmic rays. The Sun is one of the main sources of energy for maintaining our climate. We still have much to learn about the way aerosols and other gases affect regional and global climate. During the solar eclipse, the energy source decreases temporarily. Concentration of atmospheric gases, cloud, and aerosol concentration depends on the variation of the energy from the Sun. During the total solar eclipse on 23 July 2009, it was possible to correlate some atmospheric parameters like cloud cover, SO_2, NO_2, and aerosol concentration variation. The longest solar eclipse of the century cast a wide shadow for several minutes over Asia and the Pacific Ocean. Satellite data of a part of China has been selected for the study of variation of atmospheric parameters during the solar eclipse on 23 July 2009.

6.2 The Data Interpretation

Air Quality Monitoring and Forecasting in China (AMFIC) addresses atmospheric environmental monitoring over China. The system uses satellite and *in situ* air quality measurements and modeling to generate consistent air quality information over China. The data downloaded in this study cover the status of NO_2, SO_2, aerosol, and cloud cover (http://www.amfic.eu/index.php).

Extraterrestrial Remote Sensing and Climate Change, First Edition. Saumitra Mukherjee.

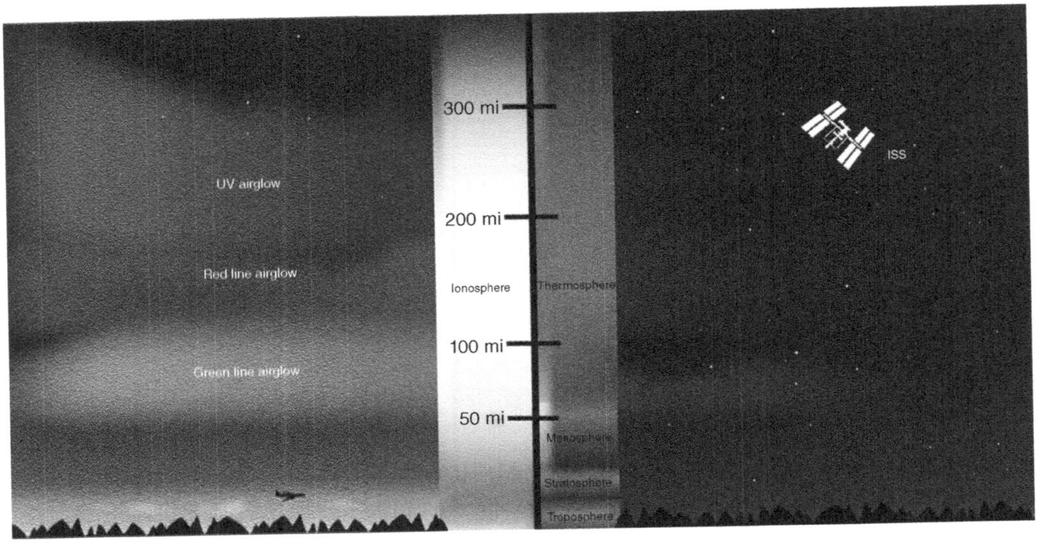

Figure 6.1 A layer of charged particles, called the ionosphere, surrounds Earth, extending from about 50 to 400 miles above the surface of the planet. Source: Credits: NASA's Goddard Space Flight Center/Duberstein.

6.2.1 Tropospheric NO_2 Data

Nitrate anomaly in different components of the environment of the Earth has been a prime concern for climate change experts. Column NO_2 retrievals with the DOAS (differential optical absorption spectroscopy) technique and the KNMI (Koninklijk Nederlands Meteorologisch Instituut) data combined modeling/retrieval/assimilation approach. The slant columns from GOME (Global Ozone Monitoring Experiment), SCIAMACHY (Scanning Imaging Absorption Spectrometer for Atmospheric Cartography), and GOME-2 observations were derived by BIRA-IASB (Belgian Institute for Space Aeronomy), the slant columns from OMI by KNMI/NASA.

Sulfur dioxide concentration in atmosphere:

Sulfur dioxide enters the atmosphere as a result of both natural phenomena and anthropogenic activities, such as coal and oil burning, and SO_2 is therefore one of the atmospheric species related to air pollution. These emissions may have an impact on air quality, and therefore, it is useful to monitor these compounds. For the Air Quality SO_2 Service, a set of geographic regions covering industrialized China has been defined and each of these regions was monitored. Concentrations of SO_2 were derived from UV measurements by satellite-based instruments, which are SCIAMACHY, OMI, and GOME-2. Recently, it was found that the Taj Mahal (made of marble) in Agra, India, is being affected by the polluted air from the Mathura refinery.

6.2.2 Aerosol Data

Aerosol mapping is considered to be of prime importance for several environmental crises including climate change, air pollution, and recent ongoing pandemics (2019–2021). Global aerosol data are retrieved from GOME and SCIAMACHY measurements by ENVISAT satellite. The retrieval is carried out by top of atmosphere (TOA) reflectances in nadir-viewing conditions, using channels selected in the most transparent atmospheric windows. The algorithms are able to retrieve aerosols over dark surfaces such as the sea's surface.

Level 1B processing GOME data has been re-calibrated using GomeCal package.

The FRESCO cloud algorithm (Fast Retrieval Scheme for Clouds from the Oxygen A-band) is a fast and robust algorithm providing cloud information from the O_2 A-band for cloud correction of ozone. FRESCO provides a consistent set of cloud products by retrieving simultaneously effective cloud fraction and cloud top pressure. The effective cloud fractions were derived by assuming that the clouds have an albedo of 0.8 and must therefore be interpreted as effective cloud fractions. Note that the derived cloud top pressures are rather insensitive to the assumed cloud albedo. FRESCO GO-v3 data version has been used to take into account the ground-pixel cloud coverage fraction. Aerosol classes have been selected among tropospheric aerosol types.

6.2.3 Cloud Data

For the retrieval of tropospheric trace gas columns, it is essential to have information on the cloud cover conditions. Both the cloud fraction and the cloud pressure are needed as input for the trace gas retrieval algorithms. The (effective) cloud fraction and cloud pressure are derived from GOME and SCIAMACHY measurements with the FRESCO cloud algorithm. The OMI cloud information is derived with the OMCLDO2 algorithm. The OMCLDO2 Level 2 data product contains the cloud fraction and cloud pressure and ancillary information produced by the OMI Cloud O2-O2 algorithm.

6.2.4 Cosmic Ray Data

Downloaded from Yangbajing International Cosmic ray laboratory. The YBJ International Cosmic Ray Observatory is located in Yangbajing valley of the Tibetan highland, a site chosen for its high altitude. This site was selected to compare the atmospheric variables of China and its surrounding area during the solar eclipse of 23 July 2009. In this observatory, the records used for the present study were mainly cosmic γ-radiation intensity at an energy threshold of few hundreds GeV by detecting small size air showers at high altitude with wide-aperture and high-duty cycle capability.

6.2.5 Electron Flux Data

Electron flux data were downloaded from Sun Heliospheric Observatory (SOHO) site through www.spaceweather.com site. The electron flux data were measured by the SOHO satellite in a continuous mode as electron flux. This electron flux plot contains the 5-minute averaged integral electron flux (electrons/cm^2-s-sr) with energies greater than or equal to 0.6 MeV and greater than or equal to 2 MeV.

During the total solar eclipse on 23 July 2009, it was possible to correlate some anomaly in atmospheric parameters like cloud cover, SO_2, NO_2, and aerosol concentration in the air of China and its surrounding areas with the cosmic ray data and electron flux data. Change in electron flux may have changed the ionization of the atmosphere, which is further interrupted by cosmic rays. These mechanisms can lead to the formation of condensation centers. But at the same time, interaction between cosmic rays and the atmosphere might change the transparency of the atmosphere as well as the atmospheric temperature. Calculations of changes of the condensation growth rate of water drops due to changes in temperature might have been performed. Attached figures shows gradual concentration of aerosol data from 20 to 23 July 2009. Aerosol concentration was low in the China-Tibet-Mongolia region, all of China, and Japan on 20 July and 21 July, and was found to be increasing in Guangzhou and Hong Kong on 23 July 2009. It is clear that there is a gradual rise of cloud cover during the same period. Cloud density was observed to be less intense over Lhasa on 20 July 2009, and was intensified over Beijing and North Korea on 23 July, the day of the solar eclipse. Corresponding figures show a slow rise of NO_2 during the same period. Nitrogen dioxide in the troposphere column was of very low density on 20 July over Mongolia and the northwestern part of China, which was dramatically increased in its highest peak over Beijing on 23 July 2009. The only exception has been observed in the case of SO_2, which shows low concentration from 20 to 23 July. Sulfur dioxide was found at its high peak on 20–22 July 2009 over Xining and the Lanzhou region of China, and was reduced on 23 July 2009.

Cosmic ray data show a peak value of 5420 on 20 July, but it reduces to 3850 on 23 July on the day of the solar eclipse. Sympathetic to the cosmic ray data, the electron flux data shows a low value on 23 July 2009 with higher values on 20–22 July 2009 and 24–26 July 2009.

The Sun, showing episodic changes during eclipse, may influence the terrestrial climate locally. Direct correlation of cosmic ray intensity and electron flux in Sun–Earth environment has been done with the aerosol, cloud cover, NO_2, and SO_2 concentration in the atmosphere. The study finds that during solar eclipse, cosmic ray intensity decreases locally. During solar eclipse, electron flux decreases. Cosmic ray intensity and electron flux have a role in deciding the aerosol concentration, cloud cover, NO_2, and SO_2 concentration in terrestrial atmosphere on a local scale. It will be essential to monitor the global cosmic ray intensity in different locations of the Earth along with the variation of electron flux on a regular basis to correlate the aerosol, cloud cover, NO_2, and SO_2 concentration during non-eclipse days also. The correlation may be useful to infer the variation of these atmospheric components with the changes in the cosmic ray intensity and electron flux anomaly. This observation can be replicated in various educational institutions for further correlation of atmospheric changes in space and time (Figures 6.2–6.19).

https://www.nasa.gov/multimedia/guidelines/index.html

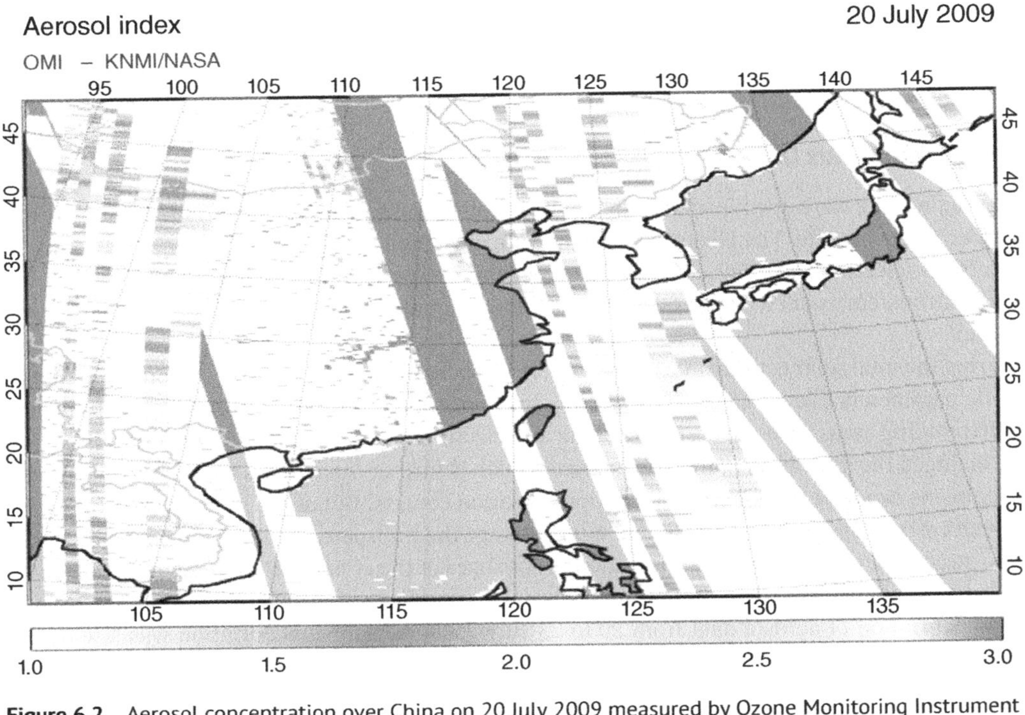

Figure 6.2 Aerosol concentration over China on 20 July 2009 measured by Ozone Monitoring Instrument showing low concentration of aerosol before the solar eclipse. Source: Adapted from KNMI NASA.

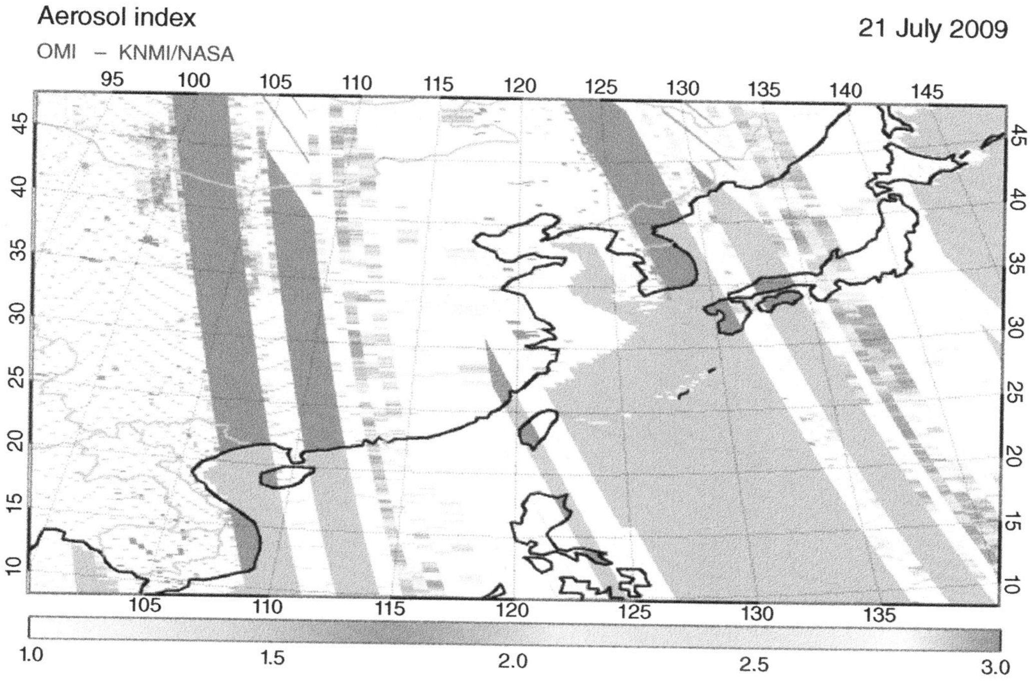

Figure 6.3 Aerosol concentration over China on 21 July 2009. The Ozone Monitoring Instrument shows slow increase of aerosol concentration before approaching eclipse. Source: Adapted from KNMI NASA.

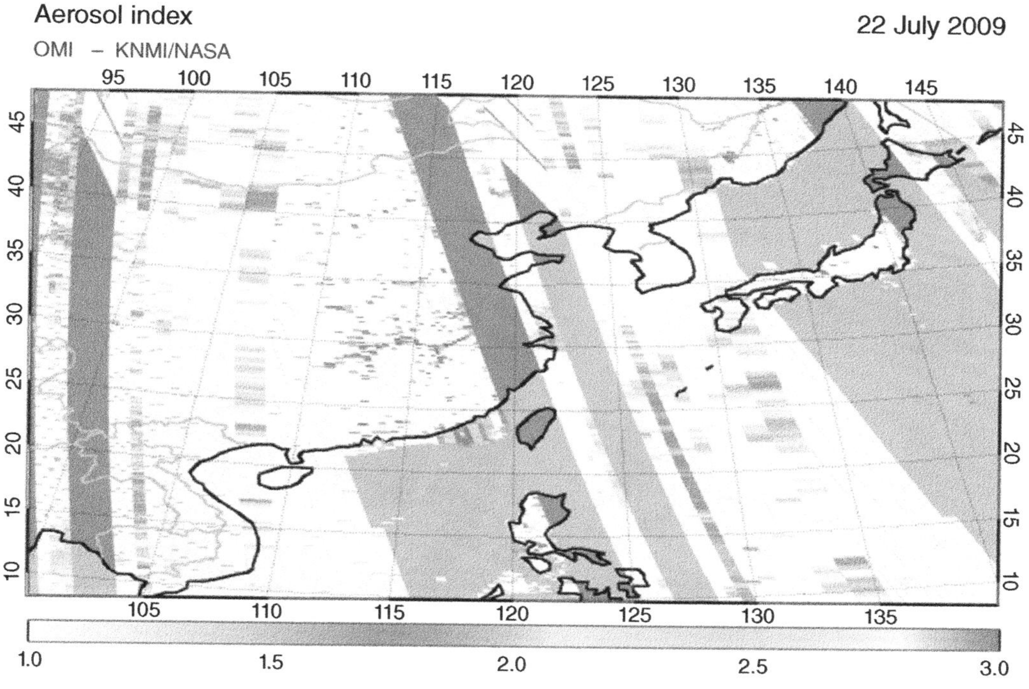

Figure 6.4 Aerosol concentration over China on 22 July 2009. The Ozone Monitoring Instrument shows concentrated aerosol formation in China, one day before approaching eclipse. Source: Adapted from KNMI NASA.

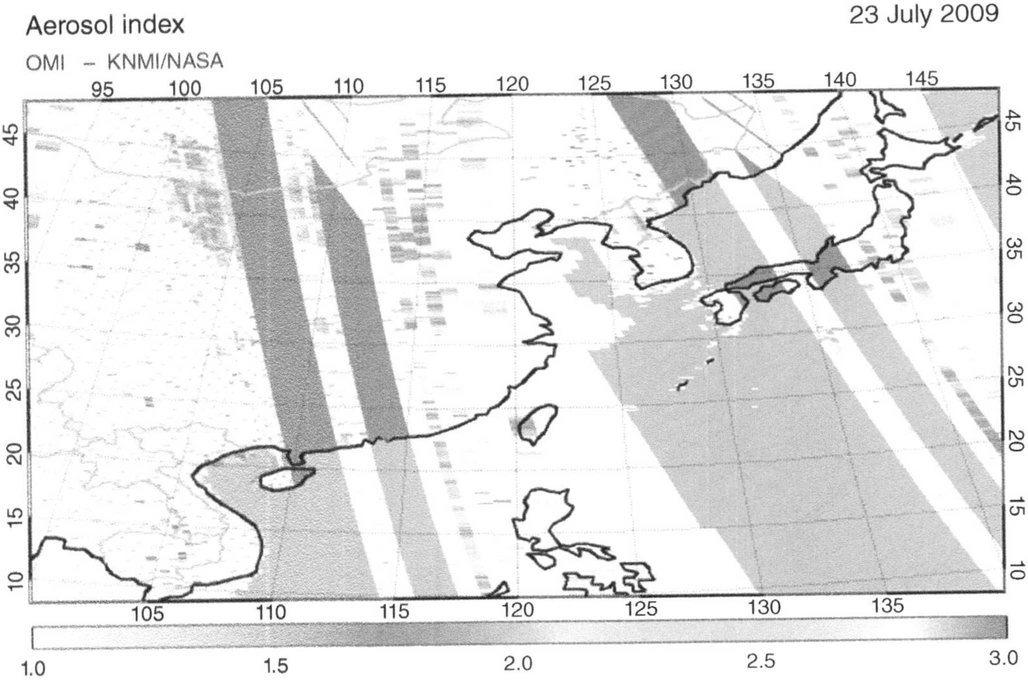

Figure 6.5 Aerosol concentration over China on the solar eclipse day, 23 July 2009. Source: Adapted from KNMI NASA.

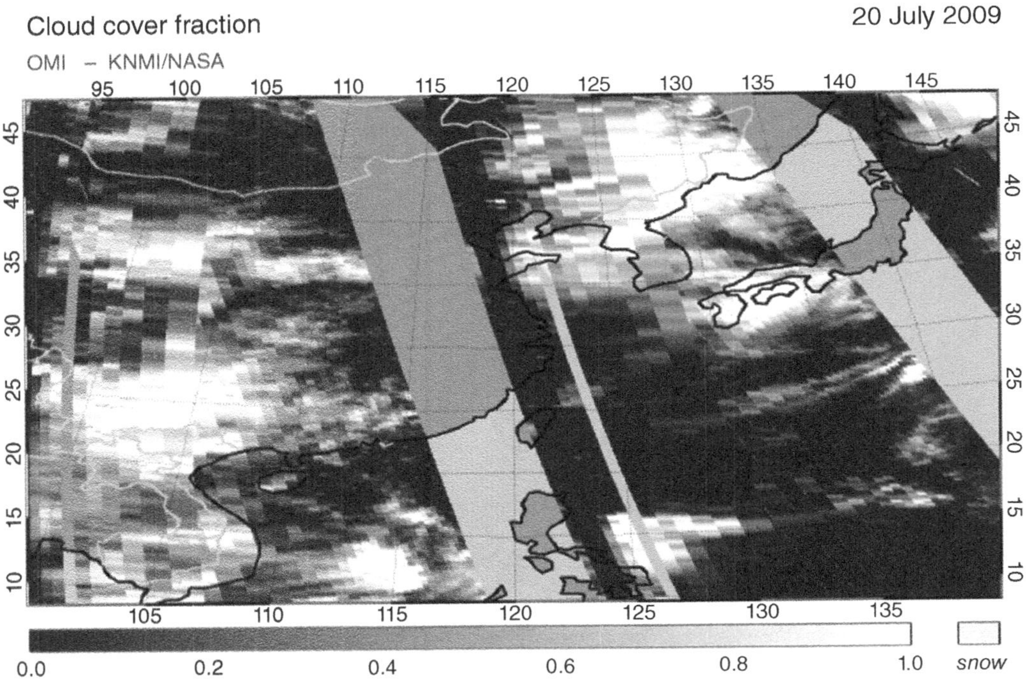

Figure 6.6 Cloud cover over China on 20 July 2009. Source: Adapted from KNMI NASA.

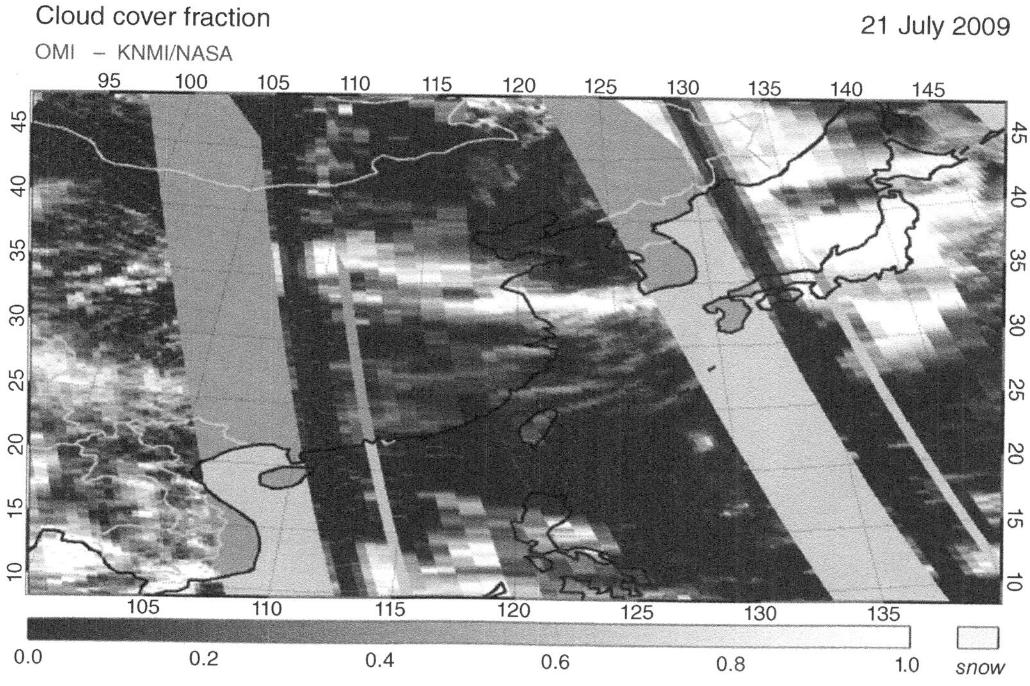

Figure 6.7 Cloud cover over China on 21 July 2009. Source: Adapted from KNMI NASA.

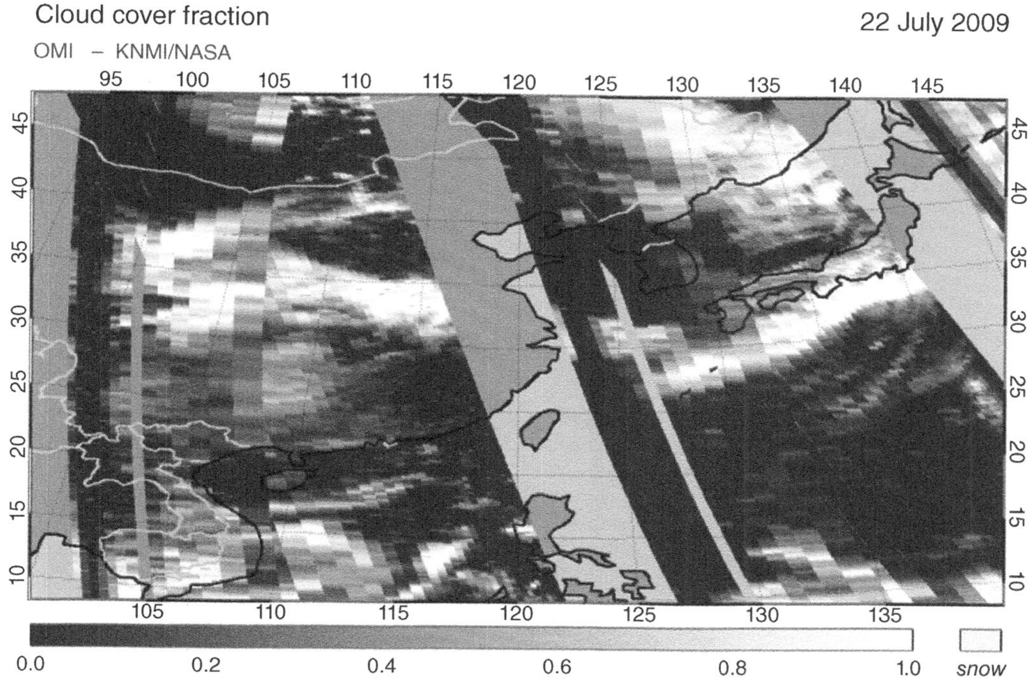

Figure 6.8 Cloud cover over China on 22 July 2009. Source: Adapted from KNMI NASA.

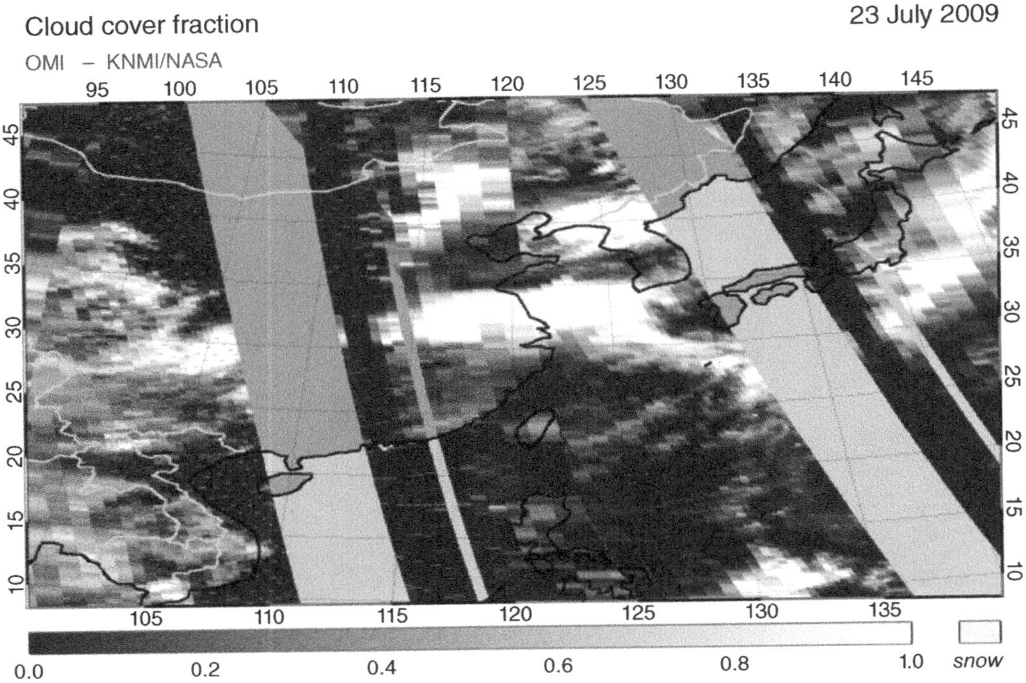

Figure 6.9 Cloud cover over China on 23 July 2009. Source: Adapted from KNMI NASA.

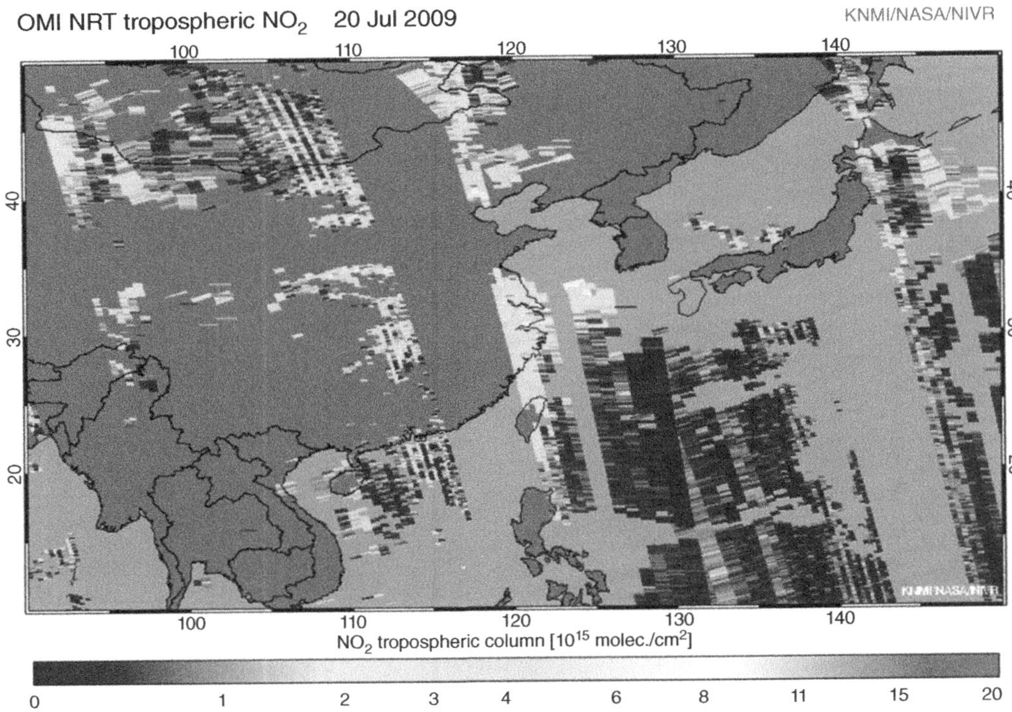

Figure 6.10 Nitrogen dioxide concentration over China on 20 July 2009. It appears to be very low in comparison with 23 July 2009. Source: Adapted from KNMI NASA.

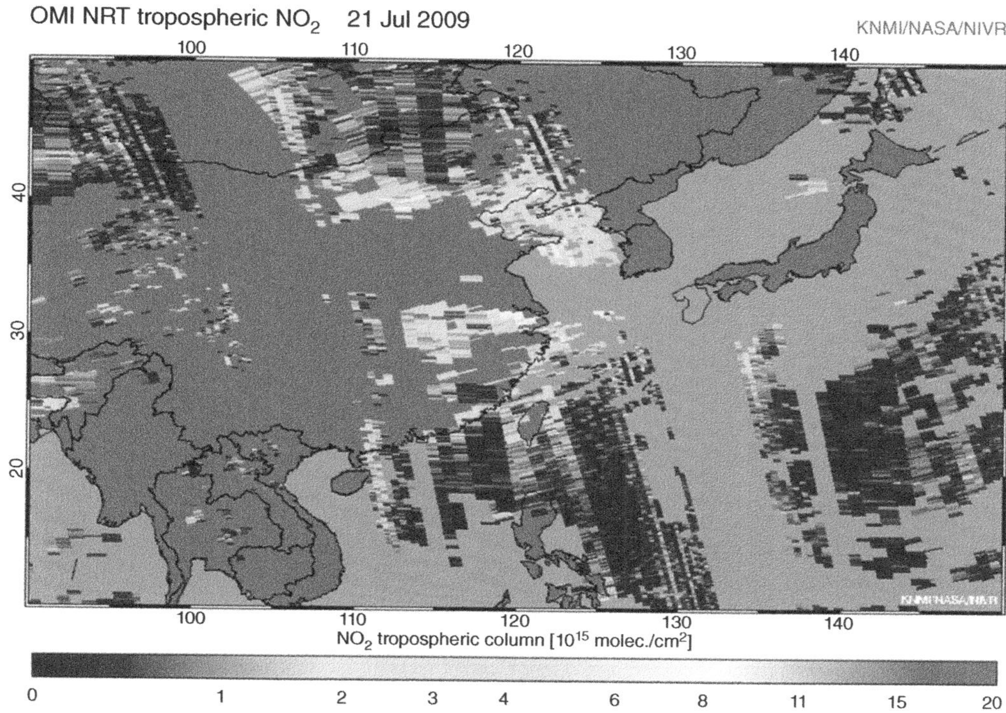

Figure 6.11 Nitrogen dioxide concentration over China on 21 July 2009, showing rising trend of NO_2. Source: Adapted from KNMI NASA.

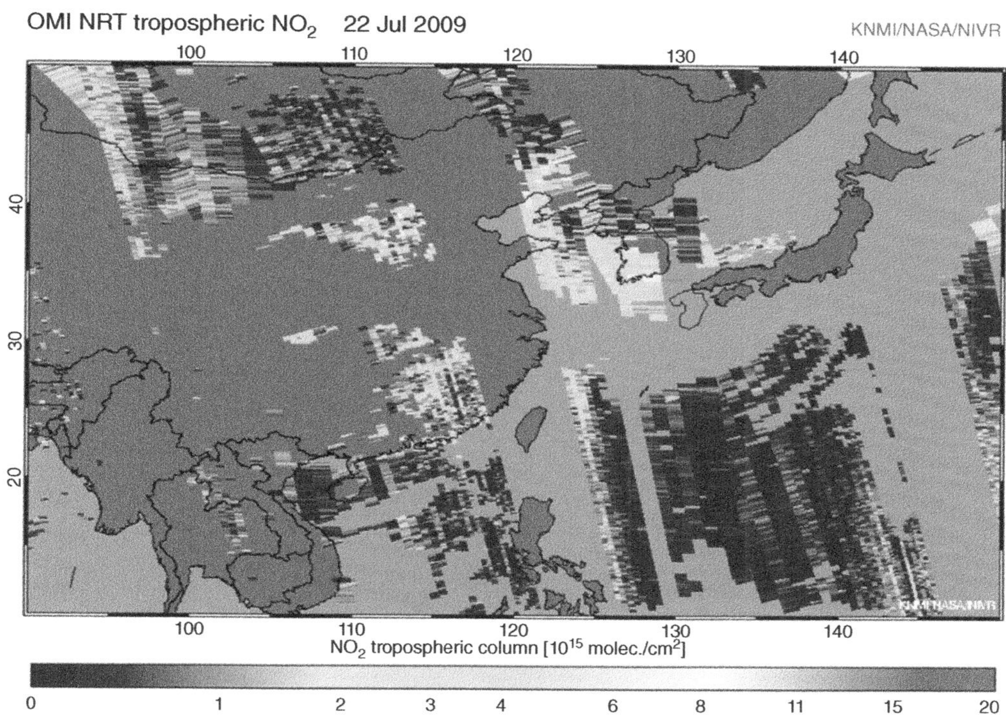

Figure 6.12 Nitrogen dioxide concentration over China on 22 July 2009, showing increasing trend. Source: Adapted from KNMI NASA.

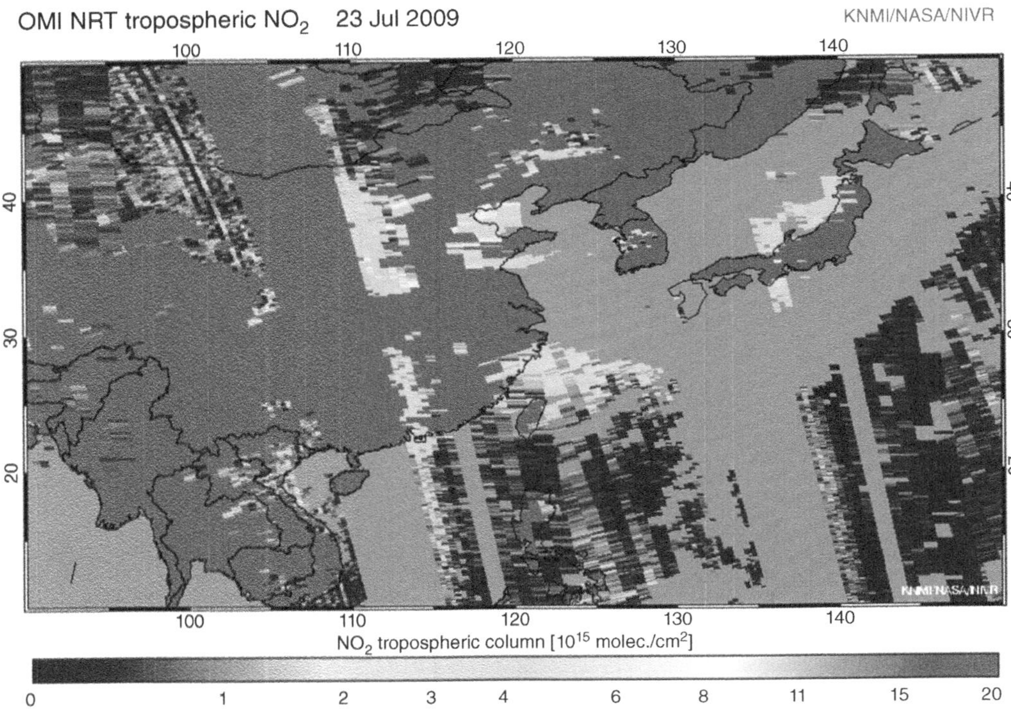

Figure 6.13 Nitrogen dioxide concentration over China on 23 July 2009. On the day of the eclipse, the concentration of NO_2 was higher. Source: Adapted from KNMI NASA.

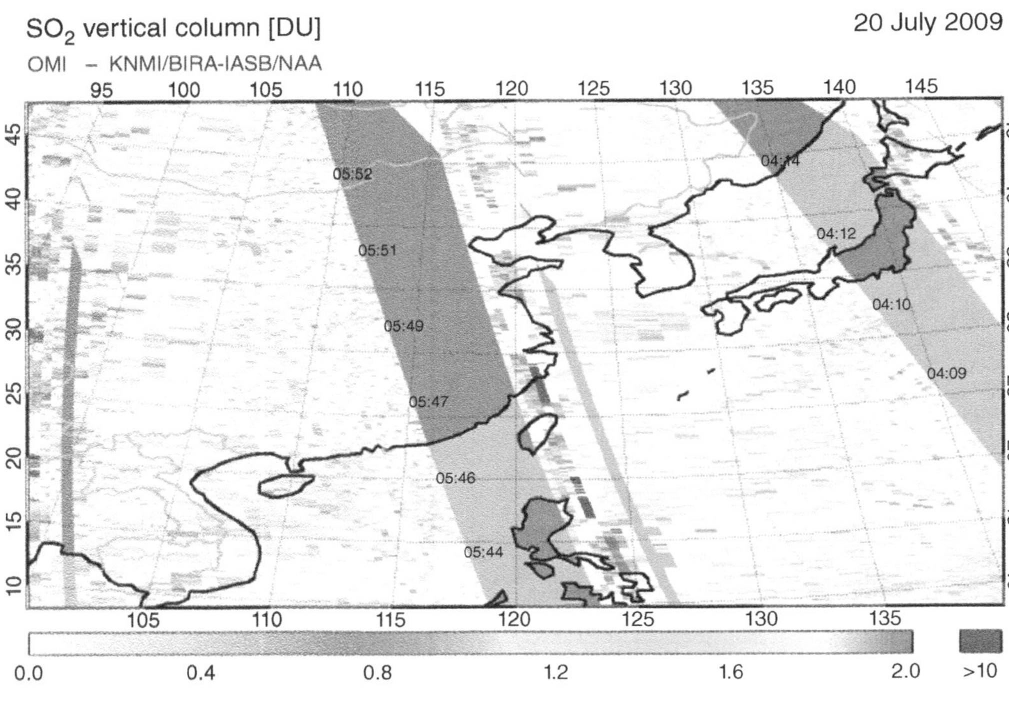

Figure 6.14 Sulfur dioxide concentration over China on 20 July 2009. Source: Adapted from KNMI NASA.

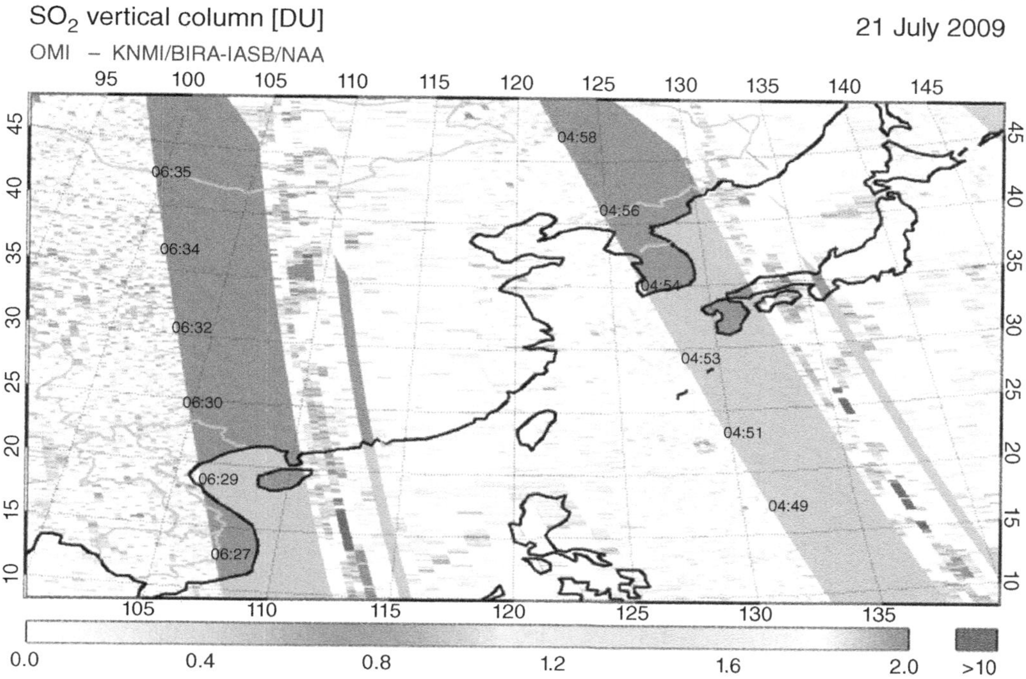

Figure 6.15 Sulfur dioxide concentration over China on 21 July 2009. Source: Adapted from KNMI NASA.

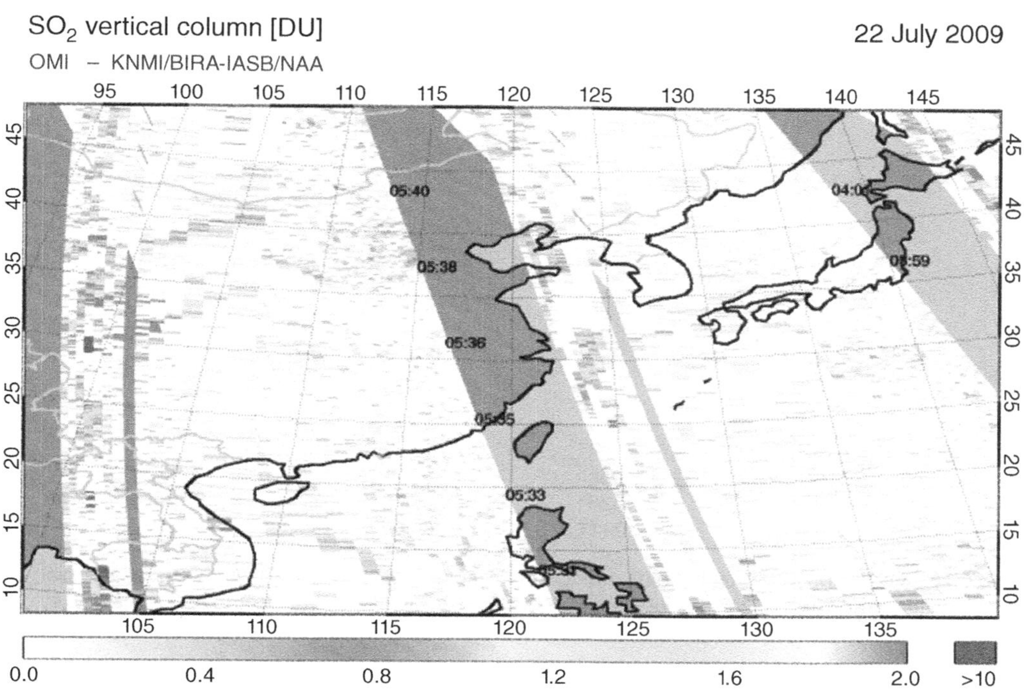

Figure 6.16 Sulfur dioxide concentration over China on 22 July 2009. It shows that SO_2 concentration has decreased before the solar eclipse. Source: Adapted from KNMI NASA.

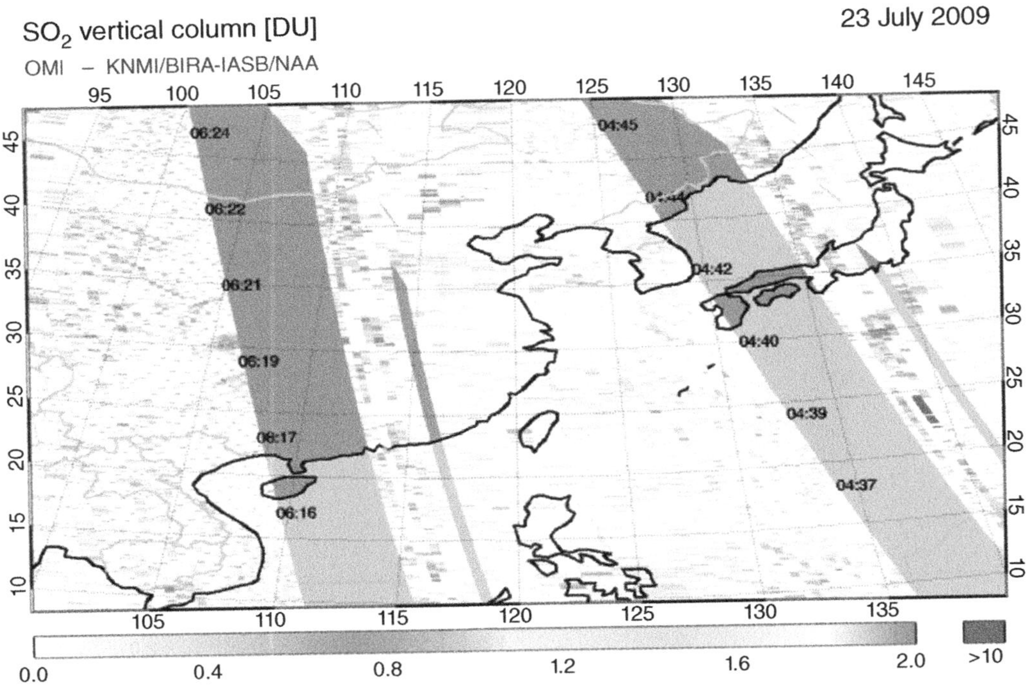

Figure 6.17 Sulfur dioxide concentration over China shows a significant decrease on 23 July 2009. Source: Adapted from KNMI NASA.

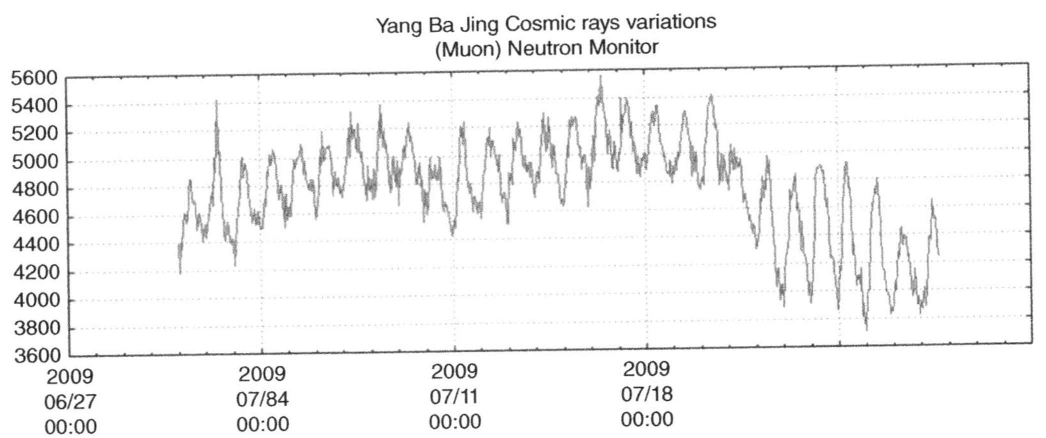

Figure 6.18 Cosmic ray variation in China during solar eclipse on 23 July 2009. It shows that during the period of solar eclipse, the cosmic ray intensity was decreased. Source: Adapted from YBJ Cosmic Ray Laboratory.

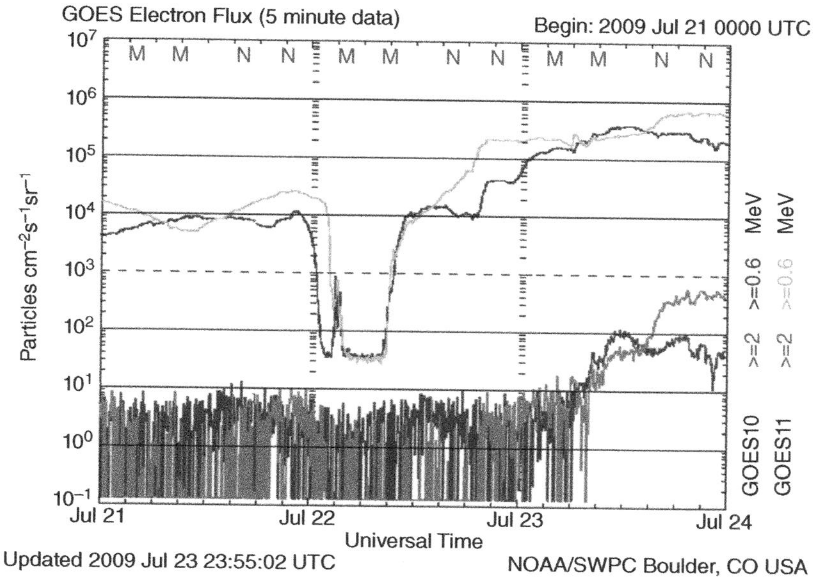

Figure 6.19 Electron flux lowering during solar eclipse on 23 July 2009. Immediately after the solar eclipse, the electron flux shows a rise. Source: Adapted from SOHO.

Further Reading

Boersma, K.F., Eskes, H.J., Veefkind, J.P. et al. (2007). Near real time retrieval of tropospheric NO2 from OMI. *Atmospheric Chemistry and Physics* 2013–2128. sref:1680-7324/acp/2007-7-2103.

Chapman, S. and Ferraro, V.C.A. (1930). A new theory of magnetic storms. *Nature* 126: 129.

Lean, J. (2000). Evolution of the Sun's spectral irradiance since the maunder minimum. *Geophysical Research Letters* 27 (16): 2425–2428.

Lean, J.B. and Bradley, R. (1995). Reconstruction of solar irradiance since 1610: Implications for climate change. *Geophysical Research Letters* 22 (1995): 3195–3198.

Marsh, N.D. and Svensmark, H. (2000). Low cloud properties 273 influenced by cosmic rays. *Physical Review Letters* 85 (23): 5004–5007.

Masmoudi, M., Chaabane, M., Medhioub, K., and Elleuch, F. (2003). Variability of aerosol optical thickness and atmospheric turbidity in Tunisia. *Atmospheric Research* 66 (3): 175–188.

Mukherjee, S. (2013). *Extraterrestrial Influence on Climate Change*. Springer. ISBN 978-81-322-07290.

Pudovkin, M.I. and Veretenenko, S.V. (1995). Cloudiness decreases associated with Forbush decreases of galactic cosmic rays. *Journal of Atmospheric and Solar-Terrestrial Physics* 57: 1349–1355.

Saumitra, M. (2008). Cosmic influence on Sun-Earth environment. *Sensors* 8: 7736–7752. https://doi.org/10.3390/s8127736 www.mdpi.com/journal/sensors.

Saumitra, M. (2013). *Extraterrestrial Influence on Climate Change*. Springer. ISBN 978-81-322-07290.

Saumitra, M. (2015). Rise in galactic Cosmic Rays before thunderstorm in Delhi. *Journal of Physical Mathematics* 6: 131. https://doi.org/10.4172/2190-0902.1000131.

Saumitra, M. (2020). *Effect of Star-Burst on Sun-Earth Environment Vision and Voyages for Planetary Science in the Decade 2013–2022*. The National Academies Press, USA. https://solarsystem.nasa.gov/studies/41/effect-of-star-burst-on-sun-earth-environment/ https://www.nap.edu/catalog/13117/

vision-and-voyages-for-planetary-science-in-the-decade-2013-2022. ISBNs: Paperback: 978-0-309-22464-2.

Saumitra, M. and Radovanović, M. (2011). Influence of the Sun in the Genesis of Tornadoes. *The IUP Journal of Earth Sciences* January: 1–16. Bochnicek, J. and Hejda, P. (2005). The winter pattern changes in association with solar geomagnetic activity. *Journal of Atmospheric and Solar–terrestrial Physics* 67 (1–2): 17–32.

Tinsley, B.A. and Dean, G.W. (1991). Apparent tropospheric response to MeV–GeVparticle flux variations: a connection via electro freezing of super cooled water in high level clouds? *Journal of Geophysical Research* 96 (1991): 22283–22296.

7

Snowfall Influenced by the Changes in the Sun

7.1 Introduction

First Potsdam Think shop of sunspot and starspot was organized by the Astrophysical Institute Potsdam, Germany, in 2002. It was a new concept, however, supported by the experts across the world (https://citeseerx.ist.psu.edu/viewdoc/download?doi=10.1.1.510.5289&rep=rep1&type=pdf). Star flares during low planetary indices (Kp) and low electron flux (E-flux) conditions of the Sun–Earth environment may have influenced the electron flux and Kp indices as well as the E-flux through the repulsion of the magnetic field in the Sun–Earth environment by star bursts. The E-flux variation found to have induced the variations in the production of ionosphere currents by free electron in the outer ionosphere periphery. These changes are produced by the geomagnetic storms originating from the Star–Sun–Earth environment changes. It was inferred that ionosphere current variations have an influence on atmospheric temperature. On 25 December 2004, hailstorms and snowstorms were were reported in large areas in the northern hemisphere; however, in lower latitudes there were drastic changes in atmospheric temperature and moisture condition. This temperature change is different in specific latitudes, longitudes, and altitudes of the Earth, as a possible effect of the solar flare would be dependent on the geomagnetic coordinates.

The Earth is in a corner of the solar system, galaxy and the universe. All members of these systems are interactive, interdependent, and interrelated. Variable solar radiation that reaches the Earth is dissipated in one of the following ways: reflection, absorption, or scattering. Initiation of climate change processes were triggered by the changes within the Sun. These changes include glacial advancement and its retreat, which has been found to be repeated in space and time. These processes are further influenced by the fluctuations of high-energy particles from the Sun as well as from galactic sources. These changes of coronal mass ejection and cosmic ray variability are also showing correlation with the components of variable environmental components. Different observatories showed muon detector and cosmic ray detector data in different latitudes and longitudes of the Earth. These datasets were correlated with the solar prominence and solar trough data of Sunspot cycle. In all correlations, it is revealed that the missing link of climate change is Forber's decrease. During these episodes, it was found that the rise in galactic cosmic rays leads to lowering of solar variables, which leads to slowdown of activities of global warming and rise in precipitation. In tropics also the measured value of solar radiation and its impact on the vegetation was observed. Space-based sensors have the potential to infer the variation in the environment of the tectonically and anthropologically active terrain of Delhi and its adjacent National Capital region. Based on the spectral response of different sensors including LANDSAT, CARTOSAT, and RISAT, it is possible to delineate the environmental stress of different environmental components.

Extraterrestrial Remote Sensing and Climate Change, First Edition. Saumitra Mukherjee.

The samples have been collected from the North Ridge Area, South Ridge Area, JNU, and some part of Haryana along the Ridge part of the Aravalli range. The sites are located on approximately 15 km wide × 70 km long transects that extended from the North Ridge Area to Faridabad. Normalized difference vegetation index (NDVI) of study area has been computed by using LANDSAT data from 2010 to 2016 in different time series. It is found that the heliophysical and cosmic ray variation has an impact on different environmental components. NDVI values range from +1 to −1 signifying the spectral response of vegetation vigor for evaluation of available soil moisture, groundwater status, and its quality in the soil as well as level of pollutants in space and time and climate variability. NDVI is dependent on not only soil and air but the influence of Sun and galactic cosmic rays were also observed by using Digital Light Meter (AG52-100) and SEVAN detectors.

The experiment shows that the solar constant varies. Sunspots and other forms of solar activity are produced by magnetic fields and their changes affect the radiation emitted by Sun, including its distribution along shorter and longer wavelengths. The Earth has a magnetic field with North and South poles. The magnetic field of the Earth is enclosed in a region surrounding the Earth called the magnetosphere. As the Earth rotates, its hot core changes the convection current pattern. This is the main reason for changes and it generates strong electric currents, which produce the magnetic field of the Earth. The magnetic field between the Sun and the Earth fluctuations affects the extraterrestrial–terrestrial changes. The magnetosphere prevents most of the particles from the Sun, carried in solar wind, from impacting the Earth. A star storm distorts the shape of the magnetosphere by compressing it at the front and causing a long tail to form on the side away from the Sun. This long tail is called the magnetotail. The star storm and Sun storm can enter through the magnetic shield and influence the atmosphere. After more than a century of controversy, solar variability effects on the climate of the Earth remains a very active research field.

7.2 Sun–Earth Cosmic Connection in Glacier and Snowfall

The middle portion of Greenland was growing and breaking from its fringes, which is responsible for the sea-level rise. Star flares during low planetary indices (Kp) and low electron flux (E-flux) conditions of the Sun–Earth environment might result in lowering further the magnetic field as well as the electron flux in the Sun–Earth environment through the repulsion of the magnetic field in the Sun–Earth environment by star flares. The E-flux variation will in turn induce variations in the production of ionosphere currents. Ionosphere currents are produced by geomagnetic storms originating from the Star–Sun–Earth environment. Can ionosphere current variations have an influence on atmospheric temperature? On 25 December 2004, hailstorms and snowstorms were reported in large areas in the northern hemisphere, while in the tropics a sudden drop of temperatures led to foggy and smoggy conditions. This temperature variation is different in different parts of the earth as a possible effect of the solar flare would be dependent on the geomagnetic coordinates. Similar phenomenon have been found in the 2005 and 2006 winters across the world. There were sudden falls in temperature and its subsequent sharp rise has been noted as an abnormal phenomenon.

Our planet is a part of our solar system, our galaxy, and the universe. All members of these systems are interactive, interdependent, and interrelated. The solar radiation that reaches the Earth is dissipated in one of the following ways: reflection, absorption, or scattering. Part of the solar radiation reaching the Earth is absorbed by the Earth and part of it is returned back. Radiation returning from the Earth is absorbed or scattered by atmosphere. The Sun is the chief driving force for terrestrial climate; the climatic variation in different parts of the Earth is determined primarily by their respective positions to the sunrays. Sunspots and Earth-directed coronal mass ejections from Sun also seem to influence the global climate. Changes in glacial deposition have been found to be in phase with changes in the orbital path of the Earth. More elusive has been a definite answer to the question of whether or not variations in the Sun's plasma emissions are capable of influencing the weather and climate at the Earth's surface. Global warming in this century has corresponded with lowered cosmic ray intensities. Cosmic rays help the formation of dense clouds in the lower atmosphere while having a small negative effect on cloud cover in the upper atmosphere. The low clouds retain more surface energy, keeping the surrounding air hot, while the high clouds reflect more sunlight into space, keeping the upper atmosphere cooler. NASA scientists have first attempted to correlate the Sun with the climate by modeling the sensitivity of the atmosphere and climate to different forcings. The solar magnetic field is the major parameter needed to reconstruct the secular variation of the cosmic ray flux impinging on the terrestrial atmosphere, since a stronger solar magnetic field more efficiently shields the Earth from cosmic rays. Further, it has been stated that cosmic rays affect the total cloud cover of the Earth and thus provide a driver for the terrestrial climate, although the physical mechanism underlying the link is still poorly understood. These point to the need for a more rigorous and thorough study of the link between Sun and climate change. In the later part of the nineteenth century, there were many claims of newfound connections between sunspots and climate. It began with the announcement by the amateur astronomer Heinrich Schwabe in 1843 that sunspots come and go in an apparently regular 11-year cycle. What followed was a flood of reported correlations, not only with local and regional weather but with crop yields, human health, and economic trends. These purported connections frequently broke down under closer statistical scrutiny, lacked the buttress of physical explanations, and were in time forgotten or abandoned. The *Chandra X-ray Observatory* has seen X-ray outbursts from a young star in McNeil's nebula, which has helped to show that its magnetic field is interacting with an orbiting disk of gas, causing it to flare up intermittently.

An attempt has been made in this study to highlight the influence of star flares on the Sun–Earth environment. Decreases in magnetic values and electron flux are also noticed after the Earth-directed star storm [1]. These geophysical parameters, e.g. the E-flux, seem to have some relationship with the snowfall in higher latitudes and higher altitudes of lower latitudes on 25 December 2004. Rainfall and the development of fog and smog on lower altitudes of lower latitude have also been recorded on the same day.

Although the Sun is known to be a variable star, its total output of radiation is often assumed to be very stable and hence its variations have negligible possible impact on climate. Testimony to this assumption is the term that has been employed for more than a century to describe the radiation in all wavelengths received from the Sun, the so-called solar *constant*, whose value at the mean Sun–Earth distance is a little over 1.37 $kW\,m^{-2}$. But in truth, the solar constant varies.

Sunspots and other forms of solar activity are produced by magnetic fields and their changes affect the radiation emitted by Sun, including its distribution along shorter and longer wavelengths. The Earth has a magnetic field with north and south poles. The magnetic field of the Earth is enclosed in a region surrounding the Earth called the magnetosphere. As the Earth rotates, its hot core generates strong electric currents, which produce the magnetic field. This field reaches 36,000 miles into space. The magnetosphere prevents most of the particles from the sun, carried in solar wind, from impacting the Earth. A star storm distorts the shape of the magnetosphere by compressing it at the front and causing a long tail to form on the side away from the Sun. This long tail is called the magnetotail. The star storm and Sun storm can enter through the magnetic shield and influence the atmosphere.

After more than a century of controversy, solar variability effects on the climate of the Earth remain a very active research field. Present work attempts to establish a new hypothesis on

Star–Sun–Earth atmospheric interactions and opens a new horizon for more accurate weather prediction research. Weakening of solar activity might have caused a mini chill in various parts of the Earth. It coincided with a large increase in the amount of beryllium-10 trapped in Greenland ice, which is an evidence of a solar flicker. This radioactive form of beryllium is produced when cosmic rays from space collide with nitrogen and oxygen atoms in the atmosphere. The magnetic field around the Earth shields the planet from cosmic rays. This field is stronger when the sun is more active, i.e. emitting more ultraviolet radiation and displaying more sunspots, hence fewer cosmic ray.

7.3 Star–Sun Influence on the Earth's Atmosphere

For hailstorm, snowstorm, or heavy cloud formation, it is essential that the Earth's atmosphere should contain enough micro aerosol particles to act as cloud condensation nuclei. Data of the star storm show that hailstorms have developed subsequently after the star storm. Since early 1992, Ulysses has been monitoring the stream of stardust flowing through our solar system. The stardust is embedded in the local galactic cloud through which the Sun is moving at a relative speed of 26 km s^{-1}. Because of this relative motion, a single dust grain takes 20 years to traverse the solar system. Observations by the DUST experiment onboard Ulysses has shown that the stream of stardust is highly affected by the Sun's magnetic field. In most of the 1990s, this field, which was drawn out deep into space by the out-flowing solar wind, kept most of the stardust out. The most recent data, collected up to the end of 2002, show that this magnetic shield has lost its protective power during the recent solar maximum. It has been reported that about three times more stardust is now able to enter the solar system [2]. The reason for the weakening of the Sun's magnetic shield is the increasing solar activity, which leads to a highly disordered field configuration. In the mid-1990s, during the last solar minimum, the Sun's magnetic field resembled a dipole field with well-defined magnetic poles (North positive, South negative), very much like the Earth. Unlike Earth, however, the Sun reverses its magnetic polarity every 11 years. The reversal always occurs during solar maximum. That's when the magnetic field is highly disordered, allowing more interstellar dust to enter the solar system. It is of interest to note that in the reversed configuration after the recent solar maximum (North negative, South positive), the interstellar dust is even channeled more efficiently toward the inner solar system. It is expected that more interstellar dust will occur from 2005 onward, but it had appeared in December 2004. An interstellar wind hit our planet. It's a helium-rich breeze from the stars, flowing into the solar system from the direction of Ophiuchus. The Sun's gravity focuses the material into a cone and Earth passes through it during the first weeks of December. Earth was inside the cone during 25 December 2004. While grains of stardust are very small, about one hundredth the diameter of a human hair, they do not directly influence the planets of the solar system. However, the dust particles move very fast and produce large numbers of fragments when they impact asteroids or comets. It is, therefore, conceivable that an increase of dust in the solar system will create more cosmic dust by collisions with asteroids and comets. We know from measurements by high-flying aircraft that around 40,000 tons of interplanetary dust enters the Earth's atmosphere each year. It is possible that the increase of stardust in the solar system will influence the amount of extraterrestrial material that rains down to Earth [3].

How the Earth's surface temperature adjusts to a given change in solar radiation depends on the processes by which the climate system responds to variations in the energy it receives. Some of these factors amplify the effects of changes that are imposed; others reduce them. Lumped together, they make up what is called the sensitivity of the climate system, which indicates the number of degrees by which the mean-surface temperature will be raised or lowered in response to a given change, up or down, in solar and/or extraterrestrial radiation or any other climate driver. To understand the impacts of star–solar variations on climate, we need to know how much the star–solar inputs vary, and how the climate system responds to these changes. The sensitivity of climate to solar radiation changes, as defined earlier, is not well known. A conservative estimate is that a 0.1% change in solar total radiation will bring about a temperature response of 0.06–0.2 °C, providing that change persists long enough for the climate system to adjust. This could take 10–100 years. Changes in the visible and infrared part of the solar spectrum alter the surface temperature by simple heating; other parts of the spectrum can also affect climate, although their paths of influence are less direct. We know, for example, that the enhanced UV radiation that pours outward from the Sun at times of high solar activity increases the amount of ozone in the stratosphere through increased dissociation of molecular oxygen. At times of minima in the 11-year cycle of the Sun, ozone is decreased. It has been also known that ozone contributes to climatic change [4]. Solar radiation received at the Earth can vary by means that are unrelated to any changes on the Sun itself. The best studied of these are very long-term changes in the Earth's orbit around the Sun, which alter the distribution of sunlight both geographically and seasonally. They are now believed to trigger the coming and going of the major Ice Ages. As such, they may provide a powerful demonstration of the impacts of changes in solar radiation on the climate system. The changes involved arise from gradual shifts in the shape and orientation of the Earth's orbit around the Sun, and in the present $23^{1/2°}$ tilt of the Earth's axis of rotation. These cyclic changes, brought about by the changing gravitational pull of the other planets and the Moon, introduce periods of about 19, 23, 41, and 100 thousand years in the distribution of sunlight over the globe. The total annual dosage, averaged over the entire surface, varies by up to 0.1%, while more specific, seasonal changes at any place can reach a few percent. Such changes are apparently sufficient to trigger major changes in climate, hence implying that the Earth's climate system may be more sensitive to small solar irradiance perturbations than one might think. Climate simulations are yet to unable the sudden rise in high cosmic ray or solar prominence. Similar periodicity is in the record of past climate change data. This long period is associated with oscillations in the eccentricity of the Earth's orbit. Changes in the Sun–Earth distance directly affect the amount of solar radiation incident on the Earth in different parts of the year. Changes in the activity of the Sun itself could exert a similar effect. Such studies of solar perturbations can serve the broader cause as diagnostic probes of the atmosphere and climate system. Ambiguities regarding projected greenhouse warming call in much the same way for clearer information regarding the role of the Sun as a possibly important contributor to the current warming trend. Climate simulations using only greenhouse gas changes predict a warming that exceeds 0.5 °C as documented in the instrumental record of the past 140 years. The reason behind difference between the observed and the predicted values may be that not all natural and anthropogenic forcing are considered in the models. If variations in the output of the Sun are indeed limited to the tenth of a percent that is recorded in direct measurements, future solar changes will likely have but a small effect, one way or the other, on the surface warming of a few degrees that is expected to result from doubled concentrations of greenhouse gases. If we consider Sun–Earth–climate connections observed in the past, we may think that star flares could potentially alter the anticipated effects of carbon dioxide and other greenhouse gases on the surface temperature of the Earth.

In January 2002, a moderately dim star in the Monoceros constellation, the Unicorn, suddenly became 600,000 times more luminous than our Sun. This made it temporarily the brightest star in the Milky Way. The light from this eruption created a unique phenomenon known as a "light echo" when it reflected off dust shells around the star. This phenomenon was followed by hailstorms in the northern hemisphere. Further, in December 2004, the star in the constellation repeated a similar phenomenon. It may be concluded that the sudden snowfall in northern hemisphere continents on 25 December 2004 has sufficient bearing on Star–Sun–Earth's atmosphere interaction. Sun–Earth environment Kp (planetary indices), proton flux, and electron flux exhibit changes. Sudden changes in these parameters may influence the environment of the Earth abruptly. If an E-flux rise is responsible for global warming, then an E-flux lowering may lead to snowfall (Figure 7.1). On 22 December 2004, a sudden fall in the electron flux was recorded by the SOHO satellite.

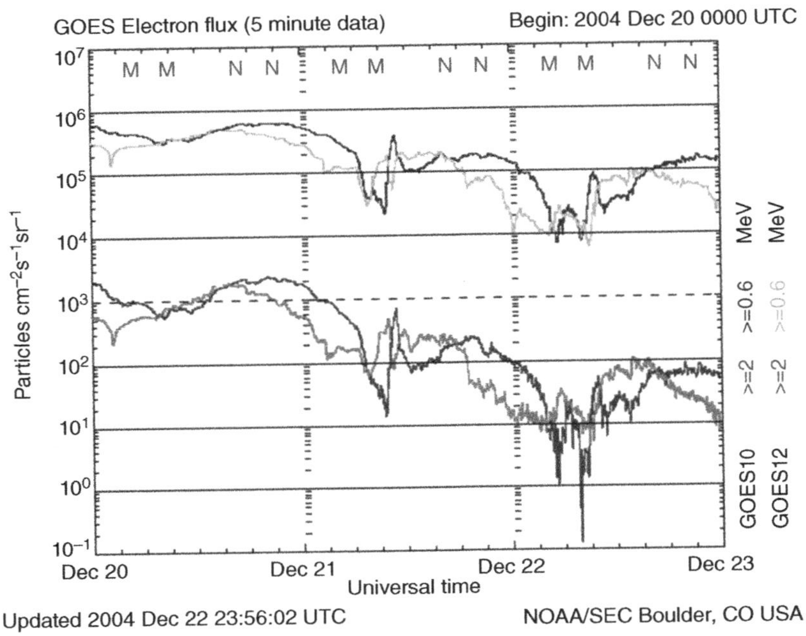

Figure 7.1 Sudden fall of electron flux 36 hours before snowfall (SOHO satellite data). Source: Adapted from NASA-ESA SOHO EIT. www.spaceweather.com.

Table 7.1 Snowfall in the United Kingdom on 25 December 2004.

Location	Snowfall Temperature (Degree Celsius)
Aberdeen	Yes 1 degree minimum
London	No
Birmingham	Yes
Manchester	Yes
Cardiff	Yes
Belfast	Yes
Crosby	Yes 0 degree minimum
Woodford	Yes 0 degree minimum

Source: Data from Meteorological Office United Kingdom and BBC weather News.

Widespread snowfall was recorded in the United Kingdom on 25 December 2005. A subsequent rise of the E-flux normalized the condition. The star flare might have influenced the E-flux and thus caused snowfall on 25 December 2004. A similar observations were noticed in other parts of the world also. Widespread snowfall was recorded in other parts of the world on 25 December 2005 and further on 23 February 2005 (Table 7.1). The temperature data of western USA shows

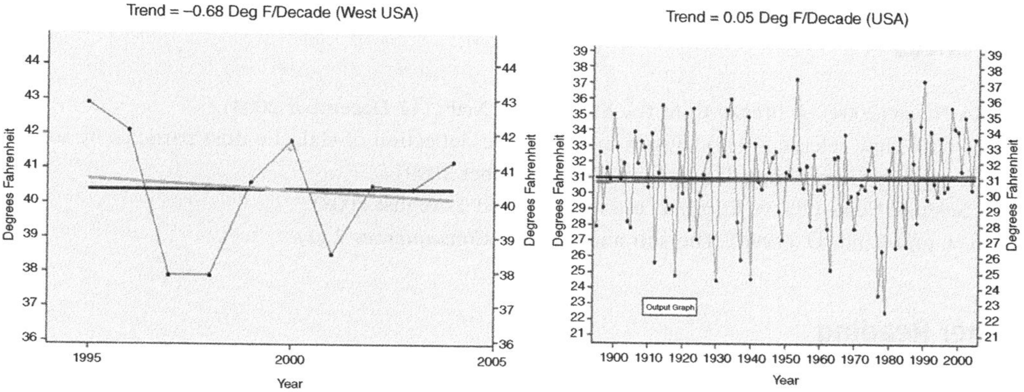

Figure 7.2 December 1995–2005 Temperature of West region, USA, shows a decline while whole USA temperature since 1895 shows rising trend. Actual, average, and trend of temperature is shown in shades of black. Source: Adapted from https://www.epa.gov/climate-indicators/climate-change-indicators-us-and-global-temperature.

Table 7.2 Sudden snowfall on 25 December 2004 and 23 February 2005.

Location	Snowfall/cold wave	Anomalous	Remarks
1. Liverpool (UK)	Snowfall	Yes	Low Kp and Low E-flux
2. Birmingham (UK)	Snowfall	Yes	-Do-
3. Manchester (UK)	Snowfall	Yes	-Do-
4. Cardiff (UK)	Snowfall	Yes	-Do-
5. Aberdeen (UK)	Snowfall	Yes	-Do-
Belfast	Snowfall	Yes	-Do-
Crossby	Snowfall	Yes	-Do-
Woodford	Snowfall	Yes	-Do-
Houghton, Michigan USA	Very high Snowfall	Yes	-Do-
Hawaii, Mauna Loa, USA	Very high Snowfall	Yes	-Do-
Boston, USA	Very high Snowfall	Yes	-Do-
New York, USA	Very high Snowfall	Yes	-Do-
Queensland, Australia	Cyclone with Cold wave	Yes	-Do-
Jammu Kashmir, Shimla, India	Snowfall with Cold wave	Yes	-Do-
Tehran, Iran	Snowfall	Yes	-Do-

Source: Data from Meteorological Office United Kingdom BBC weather News UK and NOAA USA. (After European Geological and Geophysical Sciences Letter, www.the-eggs.org)

a declining trend of −0.68 °F/decade, which is considered to be anomalous in the rising trend of 0.05 °F/decade for USA (Figure 7.2, Table 7.2). A similar phenomenon has been observed in other parts of the earth (USA and Japan) in the month of January and February 2006 followed by a star flare. Regular monitoring of star flares and their influence on the Sun–Earth environment may lead to more accurate weather prediction.

References

1 NASA News (2004). A breeze from the Star. *NASA News* (17 December 2004).
2 Max Plank New release (1999).Ulysses measures the deflection of Galactic dust particles by solar radiation. http://www.mpg.de/ (accessed 14 December 1999)
3 ESA Science News (2003). http://sci.esa.int (accessed 1 August 2003).
4 Lean, J. and Rind, D. (1996). The sun and climate. *Consequences* 2 (1).

Further Reading

Bjorck, S., Muscheler, R., Kromer, B. et al. (2001). High-resolution analyses of an early Holocene climate event may imply decreased solar forcing as an important climate trigger. *Geology* 29: 1107–1110.

Donn, W.L. (1965). *Meteorology*. New York: McGraw-Hill Book Company, Inc.

Kastner, J.H., Richmond, M., Grosso, N. et al. (2004). An X-ray outburst from the rapidly accreting young star that illuminates McNeil's nebula. *Nature* 430: 429–431.

Lassen, K. and Friis-Christensen, E. (1995). Variability of the solar cycle length during the past five centuries and the apparent association with terrestrial climate. *Journal of Atmospheric and Terrestrial Research* 57: 835–844.

Marris, E. (2006). Glacial pace picks up: Greenland's ice is breaking up at an increasing rate. *Nature* February: https://doi.org/10.1038/news060213-11.

Mukherjee, S. (2015a). Chaired a session on "Remote Sensing and Environmental Change." *International Conference on Global Environmental Change in the Himalayan Region*. Organized by German House for Research and Innovation, New Delhi (6–8 November 2015). https://mcdonnellsymposium.wustl.edu/program/symposium-program/friday-september-22-25-2016/ (accessed 22–25 September 2016).

Mukherjee, S. (2015b). Influence of solar and galactic cosmic rays on environment of the earth. *International Study for Earth-Affecting Solar Transients (ISEST)/MiniMax in México City*, México (26–30 October 2015). http://cintli.geofisica.unam.mx/congreso/ (accessed 26–30 October 2015).

Mukherjee, S. (2015c).Water Resource Management by remote sensing. *International Workshop on Water Resource Management. Perspective from Europe and South,* German House for research and Innovation, New Delhi*Asia* (28–29 September 2015). europe.uniheidelberg.de/en/newsevents/news/detail/m/heidelberg-centre-south-asia-organises-event-series.html (accessed 28–29 September 2015).

Mukherjee, S. (2016). "Water Scarcity and Agriculture in South Asia." McDonnell Scholars Academy, invited lecture on 25 September 2016 at University of Queensland, Brisbane, Australia. A session on water, food, and agriculture was chaired on 23 September 2016. https://mcdonnellsymposium.wustl.edu/program/symposium-program/friday-september-22-25-2016/ (accessed 22–25 September 2016).

Mukherjee, S. and Radovanoviæ, M. (2011). Influence of the sun in the genesis of tornadoes. *The IUP Journal of Earth Sciences* January: 1–16.

Rignot, E. and Kanagaratnam, P. (2006). *Science* 311: 986–990. https://doi.org/10.1126/science.1121381.

Solanki, S.K., Schussler, M., and Fligge, M. (2000). Evolution of the sun's large-scale magnetic field since the Maunder minimum. *Nature* 408: 445–447.

Yu, F. (2004). Formation of large NAT particles and denitrification in polar stratosphere: possible role of cosmic rays and effect of solar activity. *Atmospheric Chemistry and Physics Discussions* 4: 1037–1062.

8

Rainfall Prediction by Cosmic Ray Variability

8.1 Introduction

Rainfall prediction has achieved a new height in last decade by the advancement of terrestrial remote sensing programs. It is possible to infer the long-term prediction of rainfall and snowfall by using advanced sensors in visible infrared, thermal, and microwave regions. Accuracy can be achieved in minutes and coordinate specific. However, episodic thundershowers and catastrophic rainfall is yet to be predicted with the present-day advancement of science. The new concept of the role of cosmic rays in rainfall prediction is being emphasized in this chapter to make it a possible future research priority of meteorological sciences. The primary cosmic ray particles interact with the atmosphere and generate secondary rays, some of which will reach the surface of the Earth by reflection, refraction, and scattering mechanisms. In high-latitude regions of the Earth, where the geomagnetic cutoff is low, the lower threshold response of the neutron monitor is controlled by the atmospheric mass (about 1030 g at sea level), which limits the response threshold of the neutron monitor to primary radiation of about 430 MeV. (At the South Pole, where the surface is 2820 M above sea level, the reduced atmospheric mass lowers the primary radiation detection threshold to about 300 MeV.) Neutron monitors and cosmic ray detectors at high altitudes have higher counting rates than neutron monitors at lower altitudes because of the atmospheric absorption of the cosmic ray generated near the top of the atmosphere, which produce secondary cosmic rays.

These energetic particles are found in space and filter through our atmosphere. They come from all directions in space, and the origins are from the Sun or the galactic sources by nebular blast or by starbursts. Cosmic rays were originally discovered because of the ionization they produce in our atmosphere. Cosmic rays also have an extreme energy range of incident particles, which have allowed physicists to study aspects of their field that cannot be studied in any other way.

Initially, we named cosmic rays "galactic cosmic rays," because we did not know their origin. However, the Sun discharges a significant amount of these high-energy particles. "Solar cosmic rays" (cosmic rays from the Sun) originate in the Sun's chromospheres. Most solar cosmic ray events correlate relatively well with solar flares or coronal mass ejection.

8.2 Cosmic Rays and Rainfall

It is postulated that cosmic rays can affect the Earth by causing changes in weather. Cosmic rays can cause clouds to form in the upper atmosphere, after the particles collide with other atmospheric particles in our troposphere. The process of a cosmic ray particle colliding with particles in our atmosphere and disintegrating into smaller pions, muons, and the like is called a *cosmic ray shower*.

Extraterrestrial Remote Sensing and Climate Change, First Edition. Saumitra Mukherjee.

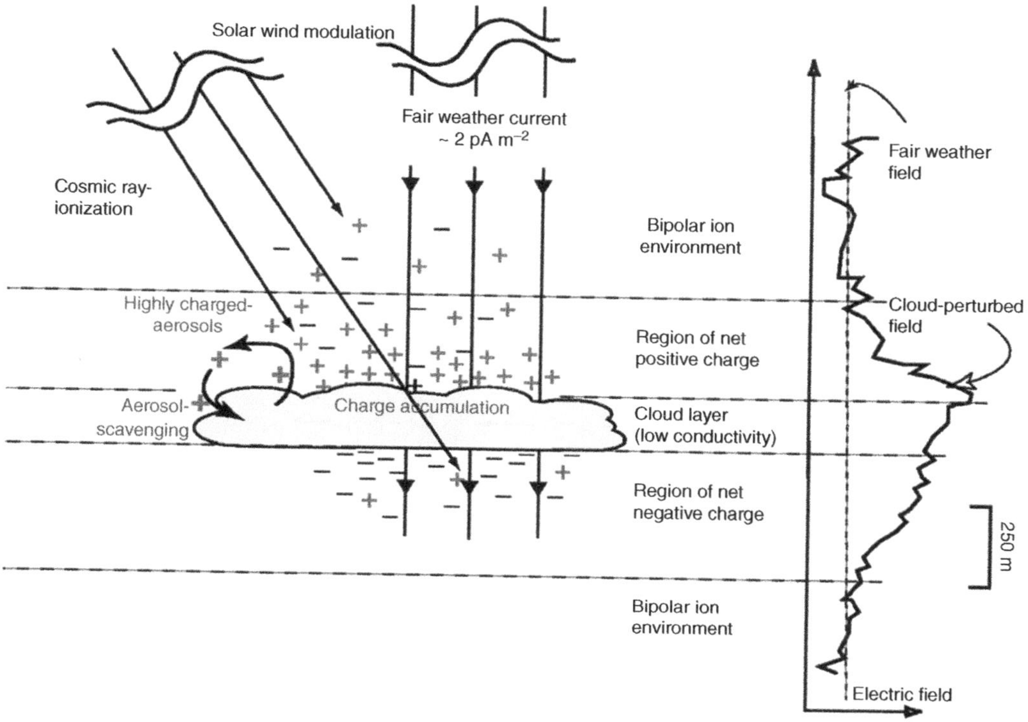

Figure 8.1 Schematic diagram of cosmic ray variability and rainfall. Source: Adapted from Carslaw et al. [1].

These particles can be measured on the Earth's surface by neutron monitors or cosmic ray detectors (Figure 8.1).

Cosmic ray flux incident on the terrestrial atmosphere and measured elementary particles on the Earth surface comprise very different entities although genetically connected with each other. Primary particle interactions with atmospheric nuclei and different methodological effects can manifest the variations of the primary flux and prevent us from understanding the dynamics of ongoing physical processes in the solar-terrestrial chain.

It is possible to recover the primary particle fluxes incident on the Earth's atmosphere; it is necessary to know the relationship between observed count rates of the detectors and the primary particles fluxes, as well as the influence of the meteorological and manmade dust or gas effects on the flux of secondary particles traveling through the lower atmosphere. It is extremely important to accurately measure the atmospheric pressure to correlate the solar modulation effects in different terrain. Manufacturing of the precise sensors equipped with automatic readout systems are the essential part of each monitoring system.

Cosmic ray intensity per unit time is sensitive to variations in the atmospheric column density above the particle detectors. This leads to a change in absorption of the secondaries and a consequent change in the counting rate, which vary in space and time. Local station pressure is a generally accepted measure for this column density. These pressure effects must be taken into account before the neutron monitor data can be used for cosmic ray studies. A period of high pressure is associated with high number of cosmic particles above the detector, and a lower detection rate results. It is possible to correlate the barometric pressure and the cosmic ray intensity with different types of pollutants in rural and urban areas. The scintillating CLOUD experiment studies how ions

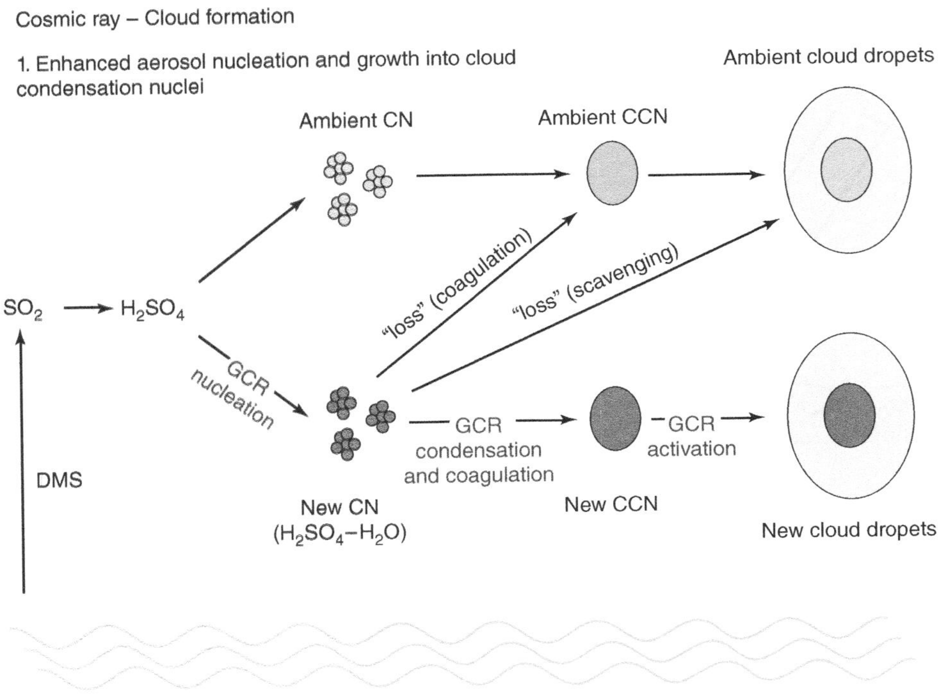

Figure 8.2 Mechanism of cloud formation by cosmic rays. (CCN = Cosmic Ray Cloud Nucleation). Source: Adapted from Ana Lopes [2].

produced by high-energy particles called cosmic rays affect aerosol particles, clouds, and the climate. It uses a special cloud chamber and a beam of particles from the Proton Synchrotron to provide an artificial source of cosmic rays. However, the cosmic rays are instead natural high-energy particles from cosmic objects such as exploding stars. It has been attempted to correlate the Monoceras Starbursts and their influence on the solar cycle, as well as on different terrestrial cycles including rainfall (Figure 8.2).

Solar beams, accelerated at the Sun, are superimposed on the uniform and isotropic cosmic ray background from galactic and extragalactic sources. Space-borne spectrometers measure the time series of the changing fluxes with excellent energy and charge resolution. High altitude and latitude surface detectors show time series of secondary particles born in cascades originated in the atmosphere by primary ions. Studies of these particles shed light on the high-energy particle acceleration mechanisms by flares and coronal mass ejection (CME) driven shocks. Time variation in the concentration of intensities of high-energy particles can provide highly cost-effective information on the key characteristics of interplanetary disturbances. Because cosmic rays are fast and have large scattering mean free paths in the solar wind, this information travels rapidly and may prove useful for space weather forecasting. Size and occurrence of the southward magnetic field component in interplanetary CMEs (ICMEs) is correlated with the modulation effects the ICME poses on the ambient population of the galactic cosmic rays during its propagation up to 1 AU. On the way to Earth (15–50 hours), the magnetic cloud and shock modulate the galactic cosmic ray (GCR) flux, making it anisotropic. Cosmic ray detector in JNU along with 14 other parts of world will be used to develop a correlation between Sun–Earth environment and environment of the Earth.

A Davis Instruments Vantage Pro 2 weather station is now installed at the Cosmic Ray Division's Aragats research station at 3 200 m (10 500 ft) altitude. It measures and sends to Cosmic Ray Division (CRD's) database (and to Davis Instrument's website) the barometric pressure, temperature, humidity, rainfall, wind speed, and direction, as well as the levels of solar and ultraviolet radiation. Meteorological information is necessary to correlate weather conditions with the generation of high-energy radiation within the Earth's atmosphere, a process called thunderstorm ground enhancement (TGE). Weather data and prediction will be helpful in scheduling the winter shift changes at CRD's high-altitude research center where winter weather conditions are often treacherous. Weather predictions are also crucial to allow visiting scientists to visit and perform research at CRD's Aragats research station.

NASA is also very interested in this research as accurate prediction and understanding of thunderstorm conditions is very important for space launches. Airlines, farmers, and the shipping industry all will benefit from this new research direction at the CRD. Prof. Ashot Chilingarian, Head of the Cosmic Ray Division and Director of the Artem Alikhanyan National Laboratory, will make a presentation about this new research at NASA's Alabama headquarters, at University of Notre Dame, and at University of Virginia.

The Weather Link Network of Davis Instruments is a collection of websites containing customer weather information that is stored on a server owned by Davis Instruments. Being part of the Weather Link Network and having a site on WeatherLink.com means that the CRD and Armenia will be a part of a community of weather researchers who use the Weather Link Network to view their own weather data and easily share this data with colleagues across the world. Armenia fills a vital gap in available weather data. Armenia's Mt. Aragats weather station is the only source of meteorological data within a radius of 1,000 km (625 miles).

The Weather Link's website is www.weatherlink.com. Visitors can click on "Weather Link Station Map" and then type "Armenia" to locate Armenia's weather station. SEVAN (Space Environment Viewing and Analysis Network) has been established in different latitudes and altitudes worldwide to infer the relation of precipitation with variable cosmic ray and other extraterrestrial data acquired from solar observation satellites. Remote Sensing Applications Laboratory of School of Environmental Sciences of Jawaharlal Nehru University in collaboration with NASA (USA) and CRD (Armenia) has developed a new instrument cosmic ray detector to monitor the environment.

SEVAN aims to improve fundamental research of the particle acceleration in the vicinity of the Sun and the space environments. The new type of particle detectors will simultaneously measure changing fluxes of most species of secondary cosmic rays, thus turning it into a powerful integrated device used for the exploration of solar modulation effects. The networks of particle detectors can predict upcoming space storms tens of minutes before the arrival of energetic protons from the Sun and hours before the arrival of huge clouds of plasma, ejected by the Sun and traveling in interplanetary space at speeds up to 2,000 km s^{-1}. Cosmic ray data from this active sensor is being used to develop a correlation between Sun and Earth to monitor the perturbations in thermosphere, ionosphere, atmospheric hydrosphere, and geosphere.

8.3 JNU, New Delhi, India, Showing Sudden Rise in Cosmic Particle Count Rate on 10 April 2012

Terrestrial as well as extraterrestrial satellite data and environmental parameter records were correlated. It has been observed that some relationship exists between the changes in environment

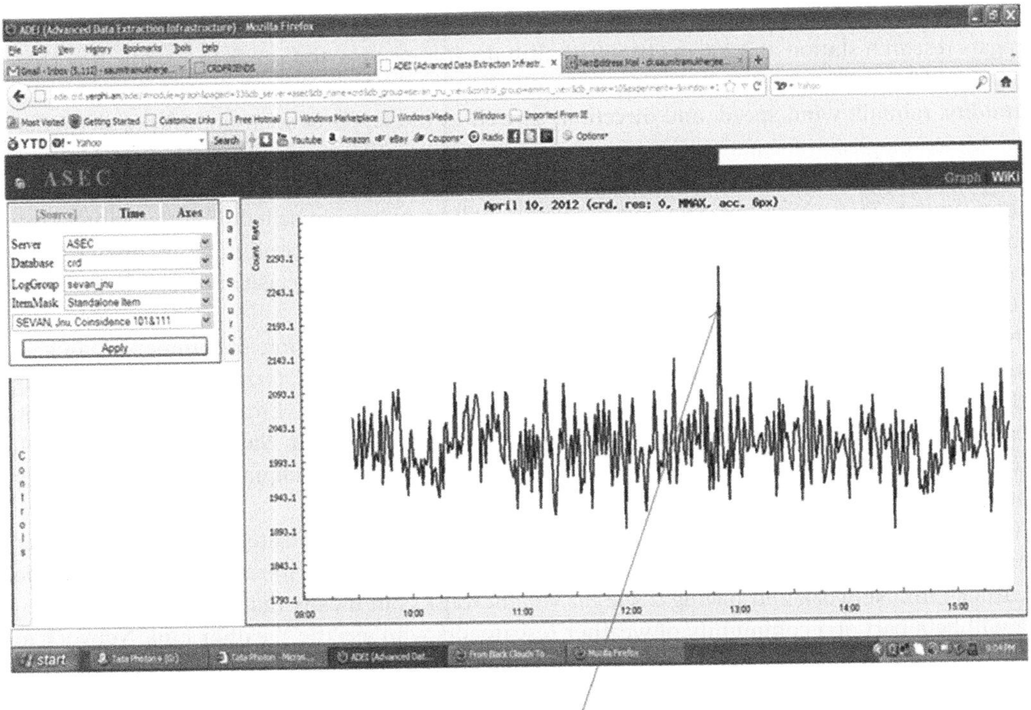

SEVAN detector at JNU showing rise in cosmic particle count rate 4 hours before thundershower on 10 April 2012

Figure 8.3 Rise in the galactic cosmic ray before thundershower in Jawaharlal Nehru University and surrounding in New Delhi, India.

and extraterrestrial phenomenon. Extragalactic cosmic rays influence planetary indices (Kp), electron flux (E-flux), and proton flux (P-flux) of Sun–Earth environment with the changes in thermosphere, ionosphere, atmosphere, and geosphere (FOREBUSH EFFECT) (Figure 8.3).

Cosmic rays on flux recently reached record levels. According to Richard Mewaldt of Caltech, "In 2009, cosmic ray intensities have increased 19% beyond anything we've seen in the past 50 years."

Cosmic rays are high-energy tiny particles traveling through space approximately with the speed of light. These particles may originate from the Sun or from the extragalactic sources. Space Environment Viewing and Analysis Network is a global instrumentation effort to detect particles from the Sun and from other distant origin. Under the auspices of United Nations, NASA (USA), and Cosmic Ray Division of Armenia, has installed this device in middle and low latitudes with the intent to improve fundamental research of space weather conditions and to provide short- and long-term forecasts of the dangerous consequences of space storms. SEVAN helps to infer changing fluxes of different species of secondary cosmic rays at different altitudes and latitudes, thus turning particle detector network into a powerful integrated device used to explore solar modulation effects. A cosmic ray detector has been installed in the Remote Sensing Applications Laboratory, School of Environmental Sciences, Jawaharlal Nehru University (JNU). This is the only cosmic ray detector at our university (or any Indian University). It will be possible to develop more research facilities in Applied Remote Sensing and Space Sciences at Jawaharlal Nehru University, New Delhi, India. With the collaboration of NASA and the United Nations, it was possible to install the international facility of cosmic ray detectors at JNU. Most of the parts of this detector were manufactured and assembled in my small laboratory. Scintillators and photomultipliers were provided by CRD, Armenia. It was possible with the initiation of AOARD and NASA to promote peaceful outer space research in developing countries. SEVAN units are installed at the Remote Sensing and Space Sciences Laboratory of Jawaharlal Nehru University, India. Similar facilities are available in Bulgaria (Moussala), Croatia, Lomnitsky Schtit, and Kosice (Slovakia). The networks of particle detectors can predict upcoming space storms tens of minutes before arrival of energetic protons from the Sun and hours before the arrival of huge clouds of plasma ejected by the Sun and traveling in interplanetary space at speeds up to 2,000 km s^{-1}. Reliable forecasts of major geomagnetic and radiation storms are of great importance because of the failures of the major space and Earth surface-based technologies, as well as posing radiation hazard to crew and passengers of satellites and aircraft and on the Global Positioning System (GPS) and other communication technologies using radio.

8.4 Influence of Solar Activity on Weather Prediction

The relationship between solar activity and change in weather is searched here by means of changes in electron flux (E-flux), proton flux (P-flux), and planetary indices (Kp) rainfall. Components of the whole universe were believed to be connected by invisible electromagnetic strings. In the electrical universe, electrons move faster than protons. Starbursts release electromagnetic anomalies, which are likely to affect the Sun as well as Sun–Earth environment. Environment of the Earth, especially movement of the clouds and its precipitation, has some bearing with changing E-flux and Kp. The Sun is responsible for developing low Kp and low E-flux conditions in the Sun–Earth environment. E-flux induces the variation of production of ionospheric currents. Ionospheric currents are produced by geomagnetic storms originating from the Star–Sun–Earth environment. Ionospheric current variation has direct influence on atmospheric temperature. On 25 December 2004 and 23 February 2005, hailstorms and snowstorms were reported in the northern hemisphere, while in the tropics a sudden drop in temperature led to foggy and smoggy conditions. This temperature variation occurs in different parts of the Earth as the effect of solar flares is dependent on geomagnetic coordinates of Earth.

On 30 May 2014, a sudden thunderstorm in New Delhi, India, was unprecedented. The wind speed was at the rate of 92–100 km h^{-1}, which claimed 23 human lives, uprooted thousands of trees, and collapsed several buildings. Routine weather prediction of different agencies was not able to predict this catastrophe.

The SEVAN detector has several sets of scintillators to record the incoming cosmic rays of different origin. The topmost 5-cm thick scintillators can record the galactic cosmic rays, while the middle unit of 25 cm thickness is capable of recording the cosmic rays from the Sun. It has been recorded that six hours before the thunderstorm in New Delhi, the topmost unit of SEVAN detected a sharp rise in cosmic rays, while the middle unit of the device recorded a lower cosmic ray intensity than normal. The pressure sensor of SEVAN has shown sympathetic rise with the 5-cm cosmic ray sensor.

Similar SEVAN units are functioning in other parts of the world, which has also shown the similar phenomena of the selective rise and fall of cosmic rays from different sources before thunderstorms.

Here, I propose a new hypothesis for the prediction of thunderstorms six to eight hours in advance. Continuous monitoring of cosmic rays from different sources and their relative abundance has the potential to help us understand the characteristics of thunderstorms. Human lives and property in different parts of the world can be saved if it is known in advance. In India, the monsoon is really erratic. Officially, until 4 September 2005, there were only 395.1 mm of rain from April 2005 compared with the normal 534.9 mm during the monsoon. The temperature recorded in New Delhi was 38 on 4 September, which was the hottest in five years. The temperature of New Delhi and its surrounding areas dropped suddenly, which led to catastrophic rainfall that continued for several days and shifted to Andhra Pradesh on 20 September 2005. Further, it can be said that sudden fall in variables in the Sun–Earth environment leads to the so-called erratic monsoon as well as erratic decrease in temperature globally.

The following Weather Model for a particular geo-coordinate can be given for the forewarning of any place on Earth:

$$\text{SSP}+\text{Kp}\,(0-9) + \text{Eflux}\,(\text{L1}-\text{M2}-\text{H3}) + \text{Pflux}\,(\text{L1}-\text{M2}-\text{H3}) + \text{CR}\,(\text{H5}-\text{L1}) = \text{SEKI}$$

where: SSP = Sunspot, Kp = Planetary indices, Eflux = Electron flux, Pflux = Proton flux CR = Cosmic ray, SEKI = Sun-Earth-Cosmic-Index for Precipitation

L = Low, M = Medium, H = High

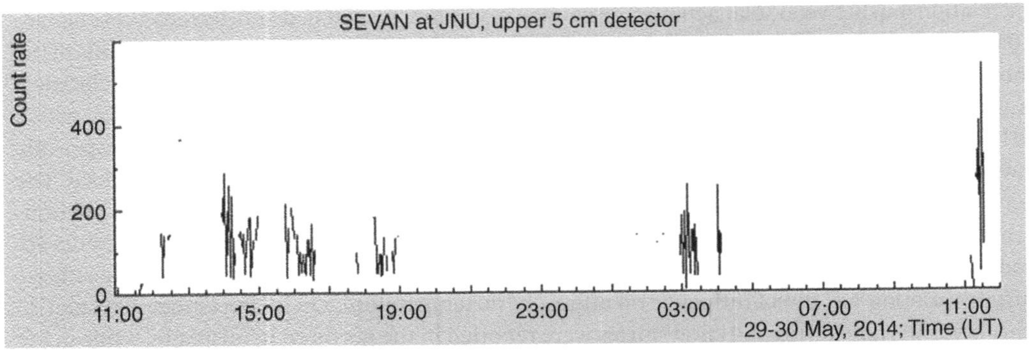

Figure 8.4 Five-centimeter thick cosmic ray sensor of SEVAN JNU showing sudden rise in count rate six hours before the thunderstorm.

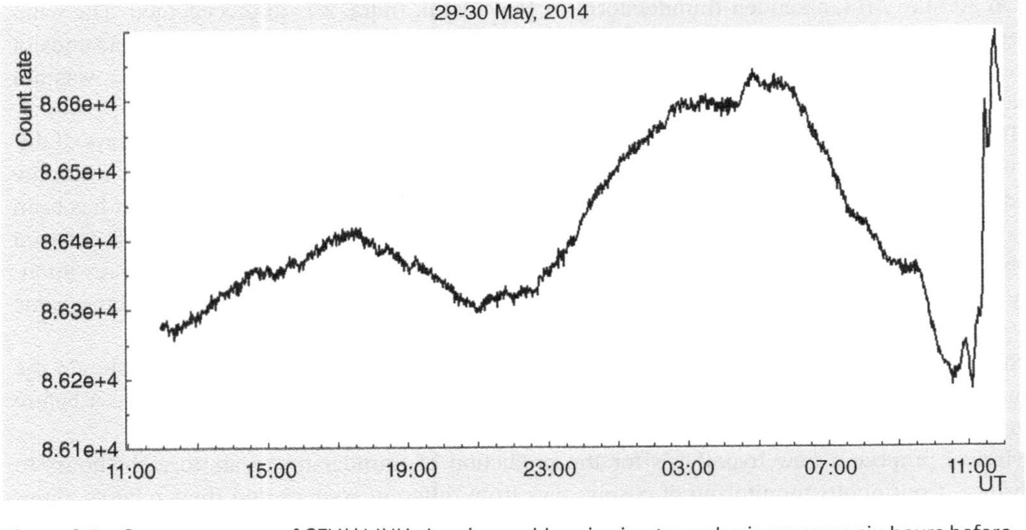

Figure 8.5 Pressure sensor of SEVAN JNU showing sudden rise in atmospheric pressure six hours before the thunderstorm.

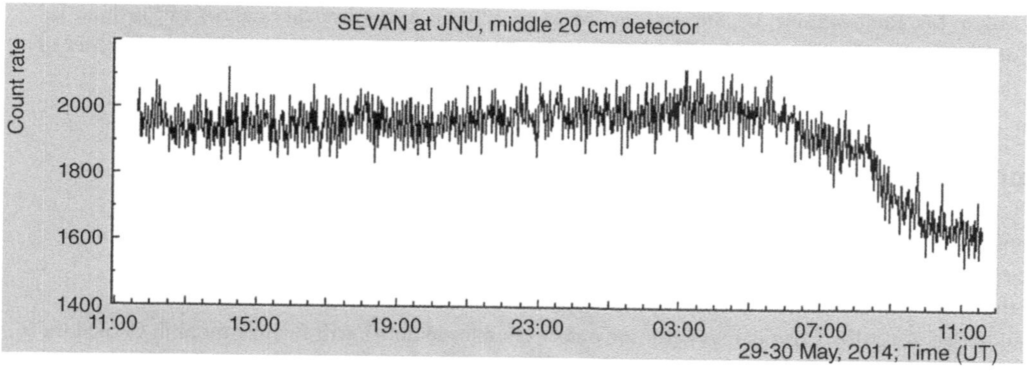

Figure 8.6 Twenty-centimeter thick cosmic ray sensor of SEVAN JNU showing sudden rise in count rate six hours before the thunderstorm.

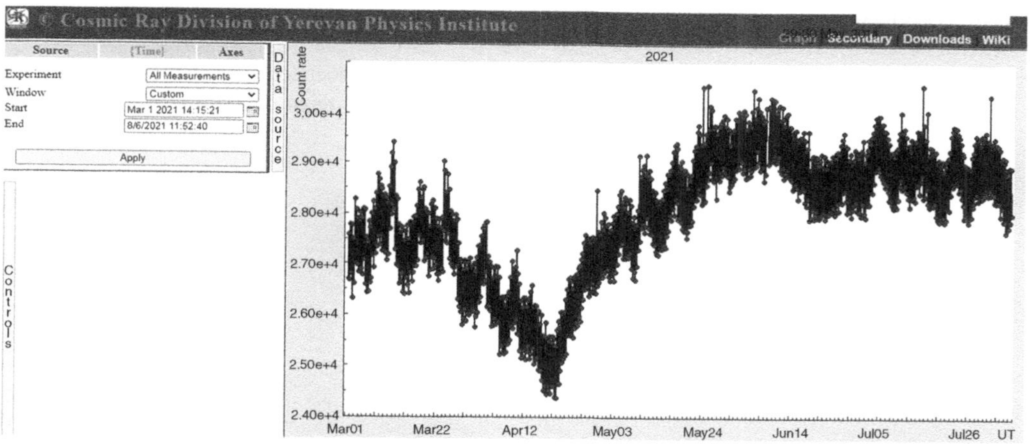

Figure 8.7 Rise in cosmic ray in the month of May 2021, which shows unprecedented rainfall.

Sudden decrease in the Sun–Earth Index leads to the assured rainfall, snowfall, or drop in temperature in the particular geo-coordinate.

Besides these, the work carried out by Milan et al. [3] shows the influence of the Sun in the Genesis of Tornadoes [3] (Figures 8.4–8.7).

The findings in JNU sevan detector has shown rise in galactic cosmic rays which is in accordance with the other networks across the world during this period.

References

1 Carslaw, K.S., Harrison, R.G., and Kirkby, J. (2002). Cosmic rays, clouds, and climate. *Science* 298 (5599): 1732–1737. https://doi.org/10.1126/science.1076964.

2 Ana Lopes (10 October 2019). From Cosmic ray to cloud. A new run of the CLOUD experiment examines the direct effect of cosmic rays on clouds. CERN Accelerating Science. Unpublished Report.

3 Milan, M., Rodovanovic, C., Stevancevic, M. et al. (2011). Astrophysical analysis of Earthquake near Karljevo (SERBIA) on 3rd November 2010. *The IUP Journal of Earth Sciences: Influence of the Sun in the Genesis of Tornadoes* 61: 1–15.

Further Reading

Babich, L.P. and Roussel-Dupré, R.A. (2007). *Journal of Geophysical Research* 112: D13303.

Dorman, L.I. and Dorman, I.V. (2005). Possible influence of cosmic rays on climate through thunderstorm clouds. *Advances in Space Research* 35: 476–483.

Mukherjee, S. (2010). Cosmic rays variation before changes in Sun-Earth Environment in Cool Stars 16, Cambridge Workshop on Cool Stars, Stellar Systems and the Sun, University of Washington, in Seattle, Washington, USA (28 August to 2 September 2010).

Mukherjee, S. (2012). Effect of climate change on water resources. In: *India Water Week 2012 Vigyan Bhawan*. New Delhi: Mineral Water Research.

Shibita, S., Munakata, K., Yasues, S. et al. (2004). Effects of atmospheric electric fields on cosmic rays. *Physical Review* 69: 123010.

Shyam, A. and Kaushik, T.C. (1999). Observation of neutron bursts associated with atmospheric lightning discharge. *Journal of Geophysical Research* 104 (A4): 6867–6686.

9

Earthquake Prediction by Using Solar and Cosmic Ray Data

9.1 Introduction

Earthquake prediction is a challenge for not only the geoscientists but the scientists of the whole world in general. The complex nature of geology mineralogy and structural geology of every kilometer of the Earth is so complex that the rock failure and its displacement estimation in space and time continue to be a big challenge. It was attempted in the past by various scientists to predict earthquakes by using geophysical and geochemical anomalies inferred by satellite remote sensing, which was partially successful in seismic microzonation, but for earthquake prediction, exact time and specific geo-coordinate pinpointed location are of paramount importance. Throughout the world, seismometers were installed to identify microtremor, induced geological, hydrogeological, and landuse/landform changes before major earthquakes. Microtremors may significantly alter the groundwater condition in an area. This change can be due to mixing of water between different aquifers, influx of water from surrounding areas, changes in dissolved gas concentrations, re-dissolving of precipitated minerals, or through infiltration of pollutants from soil or ground surface. Simultaneous landuse/landcover changes within these seismically active regions may enhance the effects on the groundwater quality and quantity, further bringing turbulence in the local aquifer system dynamics in space and time (Figure 9.1). It is well known that the Sun is the prime mover of the hydrological cycle of nature, which influences the ionosphere, atmosphere, hydrosphere, and geosphere by the changes in planetary indices (Kp), which is dependent on sunspot and solar flare activity. Within a day the Sun changes its magnetic field, which is recorded by Sun Observatory Heliospheric Observatory (SOHO) satellite.

9.2 National Capital Region, India: Tectonics and Extraterrestrial Influence

The observation of the shift of Yamuna River and changes in groundwater quality in National Capital Region of New Delhi, India, further shows the triggering of the shift of geological formation along with the aquifers that are influenced by changes within the Sun. The microtremor activities in National Capital Region of Delhi are being continuously monitored by two dedicated instruments known as Broad Band Seismometers. These instruments are placed on the Jawaharlal Nehru University (JNU) Campus, New Delhi, India, in proximity of the SEVAN network node Cosmic Ray Detector of India in the School of Environmental Sciences, JNU. These parallel data are being monitored continuously to infer the relationship of triggering of micro earthquakes and cosmic ray variability.

Extraterrestrial Remote Sensing and Climate Change, First Edition. Saumitra Mukherjee.

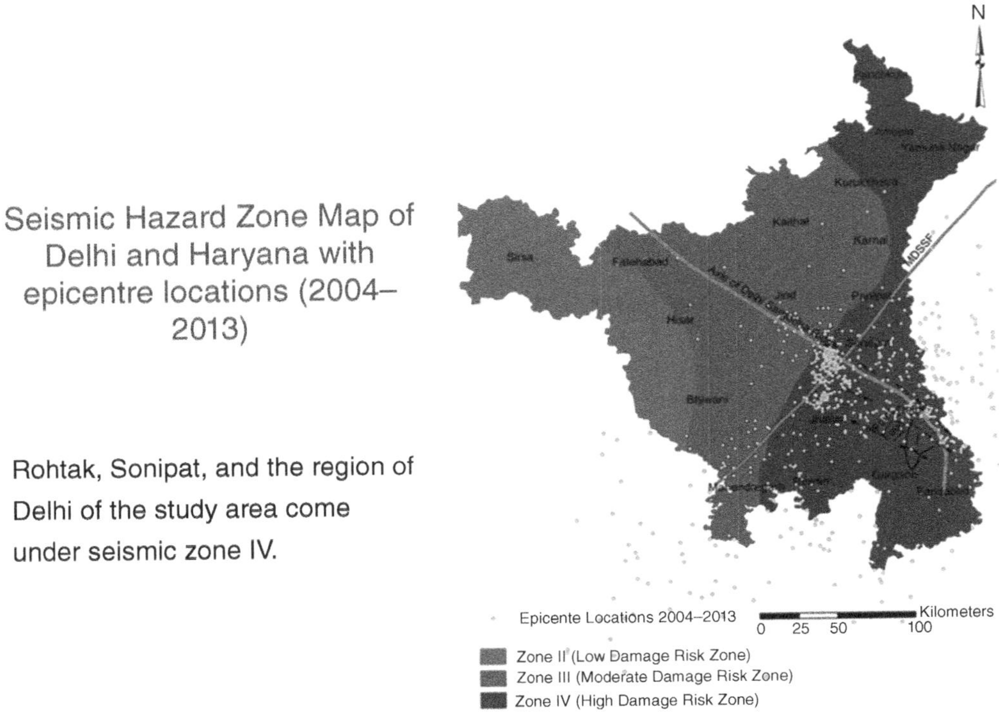

Figure 9.1 Seismically active regions of a part of National Capital Region, India, showing epicenter location of past earthquakes and microtremors.

It was found out that during extreme solar activity, the galactic cosmic rays are less numerous, while the solar proton count rises in the SEVAN detector. A new correlation established with the microtremor data is that the rise in solar proton (cosmic rays from the Sun) increases the microtremor activities. These microtremors in the New Delhi and surrounding areas create a change in the groundwater quality by releasing calcium ions from the Montmorillonite clay predominating in the Haryana adjacent to the New Delhi area.

Any phenomena that produce a change in the hydrological cycle are manifested by the changes in the terrestrial environment. Besides anthropogenic activities, the changes in the Sun and extragalactic cosmic ray intensity also influence the hydrological cycle, which is playing a major role in climate change. Solar and Heliospheric Observatory (SOHO) satellite data shows a low electron flux and planetary indices and high cosmic ray intensity in Earth-specific regions. Space Environment Viewing and Analysis Network (SEVAN) are being installed in different latitudes of the world to quantify the changes in the hydrological cycle of nature to infer climate change. Using terrestrial remote sensing (LANDSAT, IRS data) and extraterrestrial remote sensing (SOHO and cosmic ray data), it will be possible to predict the hydrological-cycle-based environmental perturbations.

Integration of seismic and hydrogeological data further shows the link in between earthquake potentiality with the groundwater quality (Figure 9.2).

The aquifer parameter data show that in earthquake-prone areas, groundwater level fluctuates, which changes not only the ionic balance of the subsurface water but affects the vegetation canopy

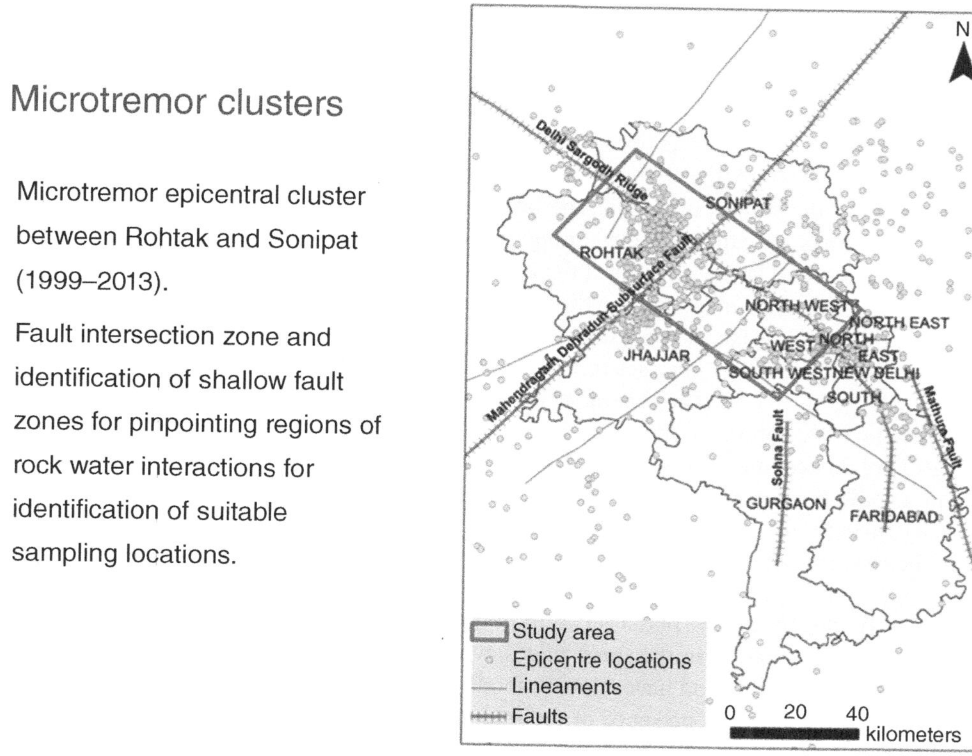

Figure 9.2 Correlation of seismically potential areas with the past earth quake location.

qualitatively as well as quantitatively. Normalized difference vegetation index (NDVI) inferred by satellite data further proves this hypothesis in earthquake-prone areas.

Earth-directed coronal mass ejection (CME) from the Sun has the potential to trigger earthquakes and change the tectonics of the Earth temporarily. The fifth state of matter (CME) acts on the geo-specific locations of the Earth. This is the reason for the presence of specific earthquake regions on Earth. Science on terrestrial tectonics is nowadays based on forms of coasts, present movements of continents, paleontology, and magnetic minerals. Earth's rotation, however, was not an interesting factor until now. A positive correlation has been inferred between rotation of the Earth, Sun, and activities of Sun–Earth environment on earthquakes and tectonics of the earth. A sudden rise in Kp values after earth-directed CME triggers earthquakes. The rise in Kp values takes place 36 hours before occurrence of the earthquake. Location of sunspots influences the geo-latitude of its mirror image on the Sun-facing part of the Earth. In hundreds of cases, it has been observed that Earth-directed coronal mass ejection (CME) is followed by changes in the Sun–Earth magnetic field, leading to remarkable changes in Earth's environment. If the CME is Earth-directed, it may produce earthquakes if active fault zones are in that latitude. It is possible to forecast the possibility of triggering of earthquakes by looking into the location of active sunspots before it harbors energy toward Earth.

Tectonic activities of the Earth were not directly linked with the rotation of the Earth. The evolution of Earth sciences and its understanding were largely dependent on paleontology, forms of the coast, and orientation of magnetic minerals. The concept of earthquake prediction was never given due credit to the effect of the Sun and its changes. Just like the Sun, the Earth has a rotation and therefore the same whirl system. It is possible to predict earthquakes in space and time if we follow the whirl pair system of the Earth, then the tectonics can be explained in a more scientific manner; however, it is not exactly the same as the Sun. The presence of ridge systems on the Earth and other planets and their moons compels us to support the varied Gyroscope whirl and Coriolis–whirl system, which explains the triggering of earthquakes very well. In comparison with Venus, Mars, and the Moon of the Earth–Sun, the Gyroscope-Coriolis forces and its effects are not the same, but they are influenced by the sun, in principle. The presence and orientation of the ridges on the Earth and different planets are capable of explaining the correlation between astronomic and terrestrial changes. The terrestrial lava whirl axes of the Earth is parallel with the rotational axis of the Earth, which bridges the gap of understanding the prediction of earthquakes.

The Atlantic ridge consists of approximately 10 ridges, which show parallel orientation with the rotational axis of the Earth. Lava rises from the radioactive part of the iron core when it arrives at the surface of the Earth or on the oceanic plate. It has been surveyed on the Earth that the axes of the whirls tend to be parallel to the rotational axis of the Earth. The lava gaps were recorded to be existing where the tectonic plates shifted from East to West and not from North to South.

The lava could flow in all possible directions quaquaversally from the point where it arrived at the surface. In contrary to this, the gyroscope law forces it in the eastern direction and the Coriolis force in the western one due to the rotation of the Earth. The two preferred directions produce lava whirls.

Interestingly, these lava flows play a very important role in stabilizing the lava gap in North–South direction with their rotation in the three ridges, which are mainly perpendicular to the equator. These whirls rotate quickly and have a low Bernoulli pressure.

The scientific origin of the presence of the mid-oceanic ridges, location is that they are at the bottom of the ocean where erupted lava can be easily cooled, where the plate is thin and the water column is greater.

It was possible to correlate the triggering of earthquakes with the heliophysical changes in the Sun. More than thousands of correlations were carried out and a few case studies are being enumerated in this piece of work.

9.3 Gujarat, Kutch, India Region: Tectonics and Extraterrestrial Influence

Active research is underway to infer the relationship between fluctuation of solar flares, sunspots, and the triggering of earthquakes. A few examples can be cited to establish the relationship between catastrophic earthquakes and variation in solar and cosmic influence. Before the occurrence of the Gujarat Kutch earthquake in India, the changes in the Sun–Earth magnetic field were observed for the first time in 2001. It was a correlation on the possibility of triggering of the earthquake by changed magnetic field between Sun–Earth environment (Figure 9.2).

A correlation was attempted with the occurrence of global earthquakes and earth-directed CMEs and changes in the magnetic field of the Earth-Sun environment (Figure 9.3). On 24 January 2001, in the Sun magnetogram inferred by Solar and Heliospheric Observatory (SOHO), an Earth-directed coronal mass was ejected from the Earth-facing side of the Sun. It took 48 hours to reach the Earth's surface. The planetary indices (Kp) recorded a sudden rise. It took two days to reach the surface of the Earth and a major earthquake of magnitude 7.9 shook the Gujarat, on the west coast of India. Throughout the world, a total of 65 earthquakes were reported on the same day. Earth-directed CMS produced a suspected invisible tail of electrified gas. The tail, which

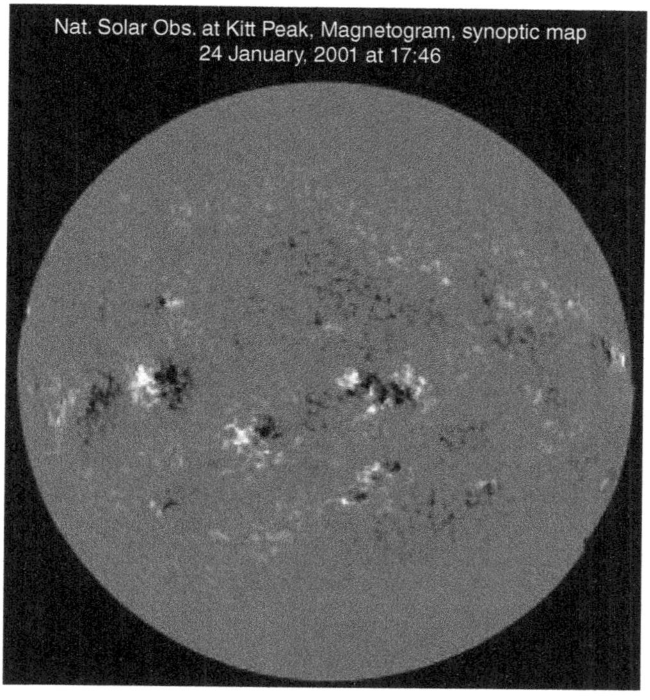

Figure 9.3 Sunspot evolving before Gujarat Kutch earthquake of 26 January 2001. Source: Courtesy of NASA.

streams from the Earth toward the Sun, was spotted by IMAGE spacecraft. Explosive events on the Sun can charge the magnetosphere with energy, generating magnetic storms that occasionally may affect the active faults in igneous geosphere to trigger the shallow focus earthquake. This correlation is observed in the case of many more earthquakes throughout the world. Besides these, it has been observed that there is an increased occurrence of earthquakes globally in this period of solar maximum. Sunspots are continuously increasing in 2001, and there are increased chances of earthquake during this period (Figure 9.4).

Similar observation was made before the Earthquake of Pakistan in 2005 and the Nepal Earthquake of 2011 (Figure 9.5).

The 25 April 2015 Earthquake of Nepal, which is also known as Gorkha Earthquake, killed 8 964 people and injured 21 952 people, according to the official record. It was a shallow focus earthquake with focus at 8.2 km below the Earth's surface (Figure 9.6).

Using Solar and Heliophysical Observatory (SOHO) satellite data, the sunspot was inferred on the Earth-facing side, which was overlaid as per the Solar-Terrestrial overlay method of Extra Terrestrial Information System (Figure 9.7)

Heat shock or cold waves, tornadoes, erratic rainfall, and snowfall are being observed before the occurrence of earthquakes. It is possible to explain the influence of stars and the Sun origin cosmic ray anomaly before earthquakes and tsunamis. Extragalactic cosmic rays measured as neutron counting rate represent an energy spectrum, which is being received by the solar system from the distant stars; the particles of cosmic rays are atom–nuclei with almost light velocity. The effects of the Sun on the environment of the Earth were found to be modulated by the geomagnetic field and the ionizing potential of the cosmic rays. Earth-directed CME and its effects on the thermosphere,

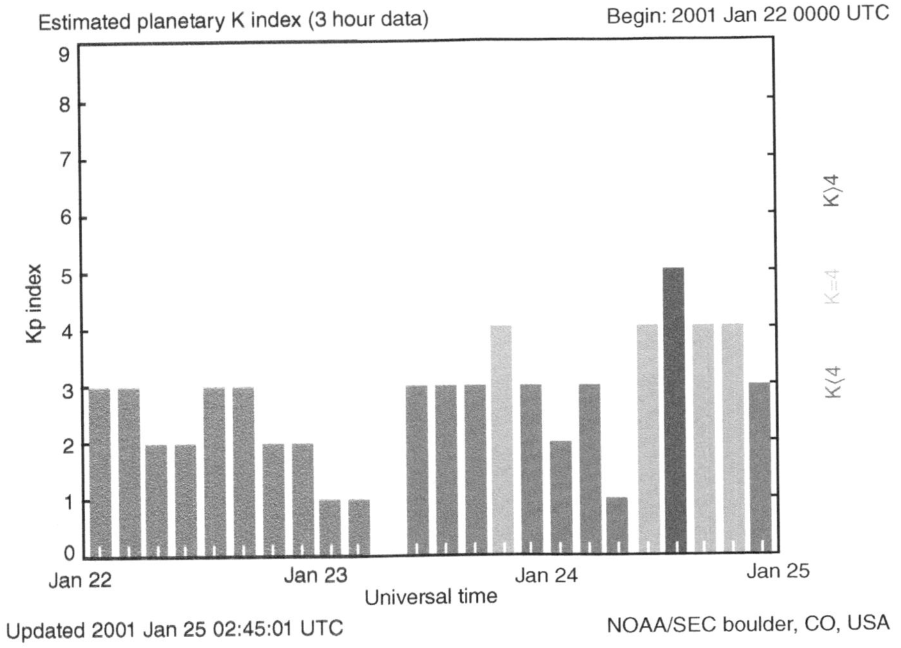

Figure 9.4 Rise in planetary indices (Kp) before the occurrence of the 7.9 Richter scale earthquake in Kutch area of Gujarat, India. Source: Courtesy NASA.

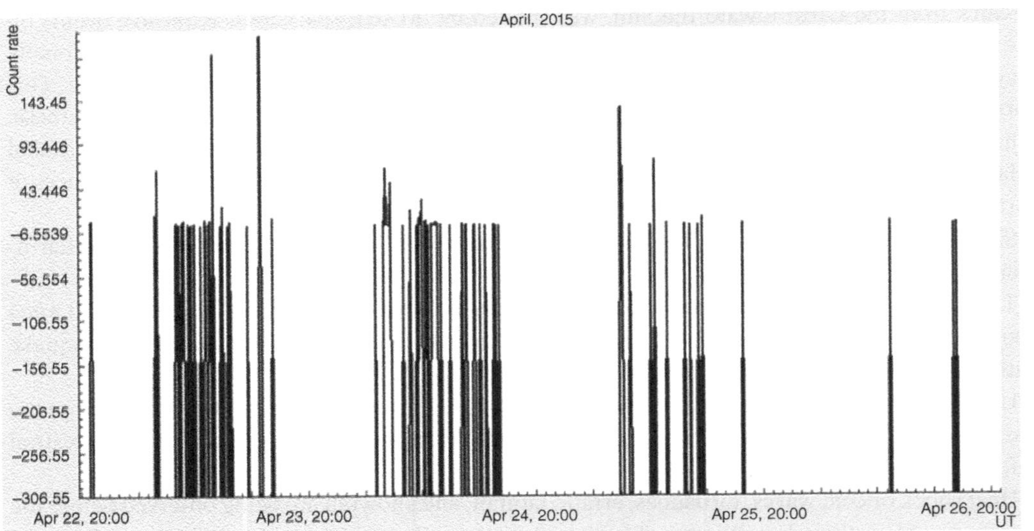

Figure 9.5 Rise in Solar Cosmic Rays before 25 April 2015 Nepal Earthquake. Source: Adapted from http://www.crdfriends.org/.

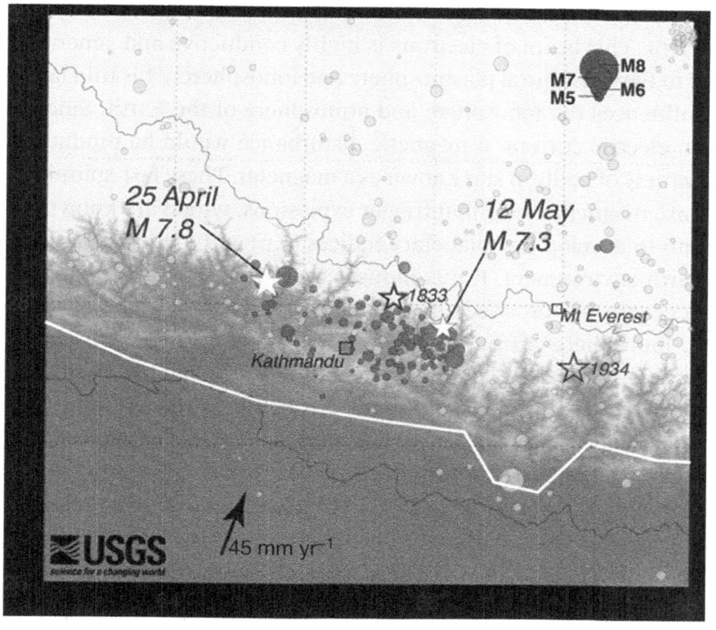

Figure 9.6 Gorkha earthquake of Nepal of 7.8 Richter scale on 25 April 2015 showing previous shocks and aftershocks. The earthquake occurred in the Indo-China plate boundary, which is shifting toward north at a rate of 45 mm yr^{-1}. Source: USGS/Public Domain.

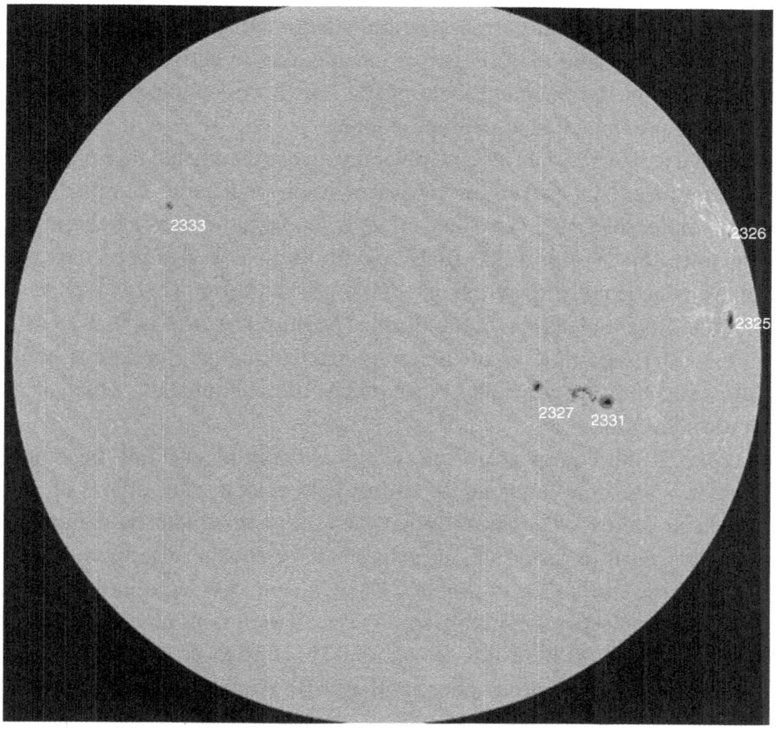

Figure 9.7 Solar and Heliophysical Observatory (SOHO) satellite data showing the Sunspot 2327 and 2331 was inferred on the Earth-facing side, which was overlaid as per the Solar-Terrestrial overlay method of Extra Terrestrial Information System.

ionosphere, and atmosphere have been studied. During Earth-directed CME, a beam of electrons (plasma) is pumped toward the Earth. This beam of electrons is highly conductive and generates an electric field that is transmitted to Earth's natural plasmosphere and ionosphere. This thin layer of charged electric field further influences the ionosphere and atmosphere of the Earth. Since a beam of electrons is carried by an electric current, a magnetic disturbance would be produced. Starbursts are caused by a special variety of neutron star known as a magnetar. These fast-spinning, compact stellar bodies create intense magnetic fields that trigger explosions, which are known as starbursts. Starbursts cause the Sun to develop low planetary indices (Kp) and low electron flux (E-flux) conditions for the Sun–Earth environment. E-flux induces the variation of production of ionosphere currents. Ionosphere currents are produced by geomagnetic storms originating from the Star–Sun–Earth environment. Ionosphere current variation has a direct influence on atmospheric temperature. On 25 December 2004 and 23 February 2005, hailstorms and snowstorms were reported in the Northern Hemisphere and higher altitudes across the world. While in the Tropics, a sudden drop in temperature led to foggy and smoggy conditions. This temperature variation was different from different parts of the Earth, as the effects of solar flares are dependent on the geomagnetic coordinate of Earth and its respective position with regards to the stars. Further, the fluctuation of atmospheric temperatures and changes in atmospheric moisture in the month of December 2005, in the first week of January 2006, and in the last week of December 2007 suggests the direct correlation of the Star–Sun–Earth environment. If the electron flux from the sun is low, with the subsequent rise in cosmic rays simultaneously anomalous snowfall and lowering of the atmospheric temperature has been observed. It would be possible to understand the movement of clouds and snowfall, as well as atmospheric moisture, if we could efficiently calculate the influence of space weather and cosmic influence on the thermosphere and atmosphere of the Earth. Based on the same hypothesis, it was found that an abnormal rise and sudden fall in E-flux, Kp index, and atmospheric temperature has the possibility of triggering earthquakes in active fault areas of the Earth due to temporary changes in the magnetic field of the Earth. The whole process was expressed as a precursor of earthquakes in active fault areas globally.

The cosmic ray detector installed in Jawaharlal Nehru University measures cosmic rays per unit time from the Sun and galactic sources. This instrument measures space environment in the form of solar and galactic cosmic rays. Increase in galactic cosmic rays has shown reduction in heliophysical parameters including Electron Flux, Proton Flux, and Planetary Indices. It has been noticed that the heliophysical as well as galactic cosmic ray intensity changes can influence the environment of the planetary system including atmosphere of the Earth. The thunderstorm of New Delhi, India, has shown the correlation with the cosmic ray and heliophysical parameters. Terrestrial environment changes manifesting as vegetation vigor can be measured by the satellite data, which also has positive correlation with the solar flux.

Extragalactic cosmic rays represent an energy spectrum, which is being received by the solar system. The particles of cosmic rays are atom–nuclei with almost light velocity. The effects of the Sun on the environment of the Earth were found to be modulated by the geomagnetic field and the ionizing potential of the cosmic rays. Earth-directed CME and its effects on the thermosphere, ionosphere, and atmosphere were studied. During Earth-directed CME, a beam of electrons (plasma) is pumped toward Earth. This beam of electrons is highly conductive. It generates an electric field that is transmitted to Earth's natural plasmosphere and ionosphere. This thin layer of charged electric field further influences the ionosphere and atmosphere of the earth. Since a beam of electrons carries an electric current, a magnetic disturbance would be produced. SOHO satellite provides heliophysical data including electron flux (E-flux), proton flux (P-flux), and planetary indices (Kp).

These data are continuously updated by NASA and being made available through the website www.spaceweather.com.

It can be concluded that a sudden l drop in Kp, E-flux, P-flux, and X-ray flux is an indication of atmospheric disturbance. This incidence was followed by anomalous behavior of atmosphere, which ultimately leads to erratic rainfall and snowfall before earthquakes at the India–Pakistan border on 8 October 2005. Due to low E-flux, the local drop in temperature in the upper part of the atmosphere leads to condensation of the clouds on the affected part of the Earth. Contrary to this, the rise in Kp indices, E-flux, and P-flux, as well as X-ray flux leads to the occurrence of earthquakes. It is possible that the high particle flux on the Earth is responsible for the release of more near-infrared radiation before the earthquake as rocks rubbing together is not responsible for the radiation field.

Further Reading

Mukherjee, S. (2001a). Space based early warning system to understand seismotectonics. *Geological Survey of India (GSI) special publication is a report of the Government of India. GSI Spl.Pub.No.65 (II)* 2001: 39–44. www.gsi.gov.in/splpub.htm.

Mukherjee, S. (2001b). Seismogenic potentiality of Delhi using remote sensing, soil geochemistry and geophysical data. *Indian Geological Congress (Spl. Vol. on Seismicity)* 2 (1): 289–297. Research Highlight in Earth System Science, DST, GOI. dst.gov.in/scientific-programme/ser-pac-es.htm (accessed 19 December 2001).

Mukherjee, S. (2006). *Earthquake Prediction*. Published by Brill Academic Publishers Koninklijke Brill NV. The Netherlands & USA: Leiden & Boston. ISBN-10: 90 6764 450 1 and ISBN-13 (i) 978 9067644 50.

Mukherjee, S. (2008). Cosmic influence on Sun-Earth environment. *Sensors* 8: 7736, 7752. doi: 10.3390/s8127736 www.mdpi.com/journal/sensors.

Mukherjee, S. (2009). Sun-Earth-cosmic connection to understand early warning of earthquakes. *Journal Earth Science India* 2 (II): 83–93.

Mukherjee, S. (2011). Sun-earth-cosmic connection to understand early warning of earthquakes. In: *Advances in Earth Sciences*, Satish Serial, Geological Survey of India 2011, xvi, vol. Vol. II (ed. S.C. Tripathi), 362. ISBN: 81-89304-92-5 http://www.earthscienceindia.info/publication.php.

Mukherjee, S. (2012). Heliophysical and cosmic ray fluctuation changes hydrological cycle. *AGU Chapman Conference on Remote Sensing of the Terrestrial Water Cycle*, Kona, Hawaii, USA (19–22 February 2012). https://www.agu.org/-/media/Files/Meetings/AGU-Chapman-Conference-Remote-Sensing-of-the-Terrestrial-Water-Cycle-Final-Program.pdf (accessed 19 December 2001).

Mukherjee, S. (2021). Changes in heliophysical parameters on Global Environment and Health. Heliophysics 2050 White Papers 2021. https://www.hou.usra.edu/meetings/helio2050/pdf/4131.pdf https://www.hou.usra.edu/meetings/helio2050/pdf/helio2050whitepapers_program.htm (accessed 19 December 2001).

Singh, P. and Mukherjee, S. (2020). Chemical signature detection of groundwater and geothermal waters for evidence of crustal deformation along fault zones. *Journal of Hydrology (Elsevier)* 582: 124459–124473.

10

Global Warming and Changes in the Sun

10.1 Introduction

Glacial melting, vapor anomaly in atmosphere, sudden tornado, storms, and tree ring variation are the manifestations of global warming. In ice cores and sediments, the fluctuation of oxygen isotopes shows that the erratic behavior of atmospheric temperature variation is not new. Quick decadal changes in temperature, radiocarbon, and its sympathetic correlation with solar flare and sunspots testify that the Sun is the major factor of climate change. The effect of Forbush decrease is likely to slow down, and it is expected that till 2040, there will be another peak of the sunspots and erratically high solar flare.

Umbrella effect can be produced by increasing cloud cover due to rise in the cloud cover, which restricts the heat emission of the Earth. Simultaneously high-energy galactic cosmic rays also interact with the atmospheric constituents and are responsible for raising the temperature of the Earth. In 2019, the pole of the Earth shifted from Canada to Siberia, which has a direct relation with the solar magnetic field change. Combined effect of rays and cloud cover led to a high atmospheric pressure in Siberia. The effect caused the East Asian winter monsoon to become stronger, researchers said.

Changes in total electron content (TEC) have bearing with the heliophysical parameter change, which influences total electron count in the ionosphere boundary. Total electron count is a measure used to characterize the conductivity of the ionosphere, which consists of ionized layers of the upper atmosphere. Free electrons in the ionosphere affect the transmission of radio waves by absorbing and reflecting, which slows the propagation of radio signals through the ionosphere. Estimates of the TEC can be used to correct transmission delays in GPS signals, which can be incorporated in GPS receivers for more accurate location estimates.

In the darkness of space between the Sun and Earth, electromagnetic radiation from radio waves to gamma rays as well as high-energy charged particles stream constantly toward the Earth. Meanwhile, in the rarefied upper atmosphere of the Earth far above the highest clouds, magnetic storms rage and howling winds drive atmospheric currents around the globe. Conditions in the Earth's upper atmosphere and further out into space, driven by powerful emissions from the Sun, are referred to as "space weather."

Global warming, melting of the glaciers, and climate change have substantial impacts of extraterrestrial perturbation during irregular solar heliophysical phenomena. Ganga River originates from the Gangetic glacier of western Himalayas in the Indian state of Uttarakhand and flows south and east through the Gangetic plain of India. It eventually empties into the Bay of Bengal. Disturbances like flash floods, landslides, and earthquakes are associated with the upper part of the river. The area is also prone to very high intensive earthquakes, which generate a number of faults as well as

Extraterrestrial Remote Sensing and Climate Change, First Edition. Saumitra Mukherjee.

interconnected lineaments. All these are changing the course of the river and its other depositional features. Various fluvial anomalies such as gradient reversals in the longitudinal profile of the channels and sinuosity variations (the river taking a turn) are taking place. The climate change induced by anthropogenic activities in the Indo-China region has accumulated aerosol. India is always threatened by its neighboring countries including China. China's industries are generating aerosol and they are accumulating in the Himalayan areas. Along with aerosol, water vapor and rise in galactic cosmic rays from time to time generates clouds that bring abnormal torrential rains in the Himalayan foothill zones, which melts the glaciers abnormally. A case study was conducted in the Kedarnath region by JNU New Delhi-based Space Environment Viewing and Analysis Network (SEVAN). Based on the speciation of cosmic rays it is possible to infer its source. One can now actually identify which cosmic rays are coming from the sun and which from galactic sources. The hypothesis is being postulated for the first time on the mechanism of the heat transfer from the charged protons to the upper and lower atmosphere of the Earth. The heat wave (radiation) that came from the sun was stored in the Van Allen belt. When that energy was released, it affected the atmospheric temperature suddenly and raised it to a great extent. Due to this, reservoirs of all sizes went through an excessive evaporation in the Himalayan region. Along with that there was a very high extraterrestrial cosmic ray anomaly, which was measured in our laboratories with equipment manufactured at Jawaharlal Nehru University with the funding support of NASA. The heat from the sun is intermittently being captured in the Van Allen belt, which further accelerated the protons. The primary effect of heat transfer initiated by the proton flux from the sun was responsible for the rise in atmospheric temperature. This interfered with the transfer of hydroxyl ions between the atmospheric water molecules and the aerosol particles that were already in existence due to release from glacial melting. They concentrated and became a nucleus for the cloud formation. On one side, there were the aerosol particles, and on the other, cosmic rays and high level of water vapor; all three together combined to form the clouds that gave the torrential rains in Himalayan mountain range. According to new research, the conclusion is that unplanned human activities in the Himalayan terrain coupled with some unusual extraterrestrial events triggered the massive cloudburst over Kedarnath, resulting in the catastrophic flood in the Uttarakhand region every year. Large numbers of people visit the area for either worship or tourism, so it is important to understand the region in detail and record the changes that take place there, which will enable some precautionary measures.

10.2 Impact of Solar Changes on Global Warming

Although the extended solar minimum is continuing from 2009, occasional solar flare pours an excessive amount of heat in the form of proton flux on the planetary systems including the Earth. "The rise in proton flux – an unusual solar event where protons emitted by the Sun are accelerated to very high energies – was responsible for the anomalous rise in atmospheric temperature. High concentration of aerosol trapped in the atmosphere and glaciers in Indo-China border initiated a nucleation process in the concentrated water vapor to trigger formation of clouds for the torrential rain and cloud burst on 16 and 17 June 2013," Solar and Heliospheric Observatory (SOHO) satellite recorded a steep rise in solar proton flux above 10 MeV for 12 days from May 15 to 26. During the same period, the cosmic ray intensity was also recorded at an all-time high at New Delhi's Jawaharlal Nehru University (JNU), which represents the regional cosmic ray data in the SEVAN of the Asian office of Aerospace Research (Figure 10.1).

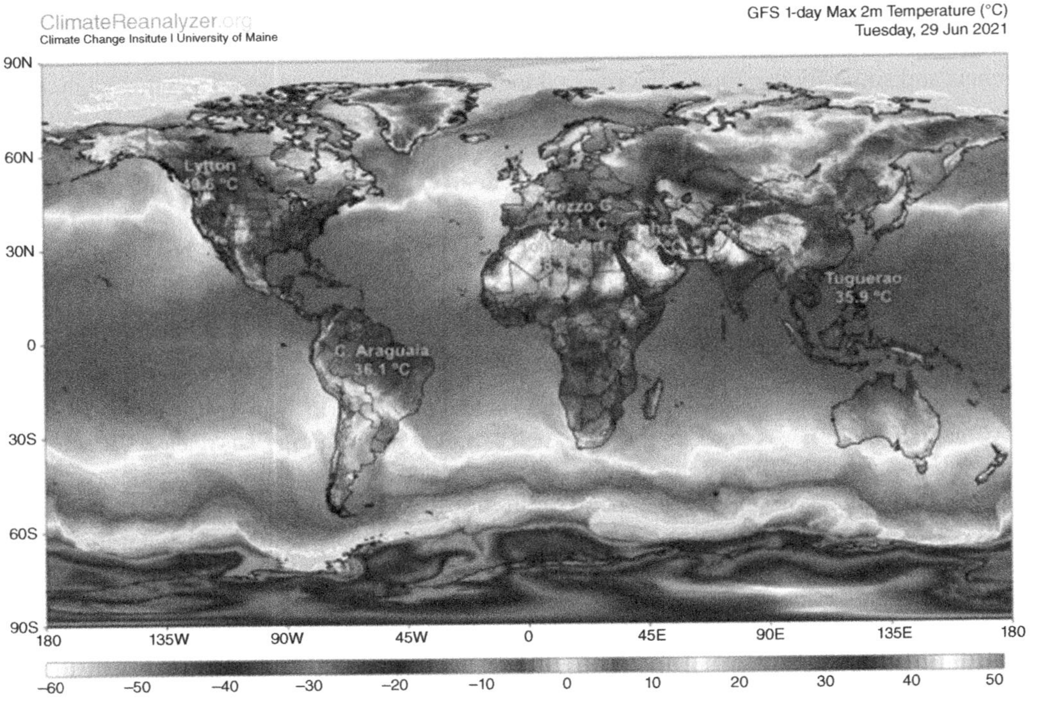

Figure 10.1 Rise in global temperature in 2021. Source: Climate Reanalyzer.org.)

Solar proton flux has been studied for the first time to explain the unusual global warming. In May 2015 (Figure 10.2), high-energy proton particles are trapped in the Van Allen belts, which is released during restoration of ionosphere and Sun–Earth magnetic field. Besides solar proton, the total electron count is also an important component to understand global warming. Total electron count is a measure used to characterize the conductivity of the ionosphere, which consists of ionized layers of the upper atmosphere. Free electrons in the ionosphere affect the transmission of radio waves by absorbing and reflecting, which slows the propagation of radio signals through the ionosphere. Estimates of the TEC can be used to correct for transmission delays in GPS signals, which can be incorporated in GPS receivers for more accurate location estimates.

The TEC is the total number of electrons present along a path between a radio transmitter and receiver. Radio waves are affected by the presence of electrons. The more electrons in the path of the radio wave, the more the radio signal will be affected. For ground to satellite communication and satellite navigation, TEC is a good parameter to monitor for possible space weather impact influenced by the changing Sun.

The TEC in the ionosphere is modified by the changing Sun. Solar minimum to solar maximum change influences solar Extreme Ultra-Violet radiation, geomagnetic storms, and the atmospheric waves that propagate from the lower atmosphere (Figure 10.3). The TEC depends on local time, latitude, longitude, season, geomagnetic conditions, solar cycle and activity, and troposphere conditions. The propagation of radio waves is affected by the ionosphere. The velocity of radio waves changes when the signal passes through the electrons in the ionosphere. The total delay suffered by a radio wave propagating through the ionosphere depends both on the frequency of the radio wave and the TEC between the transmitter and the receiver. At some frequencies, the radio waves pass through the ionosphere. At other frequencies, the waves are reflected by the ionosphere.

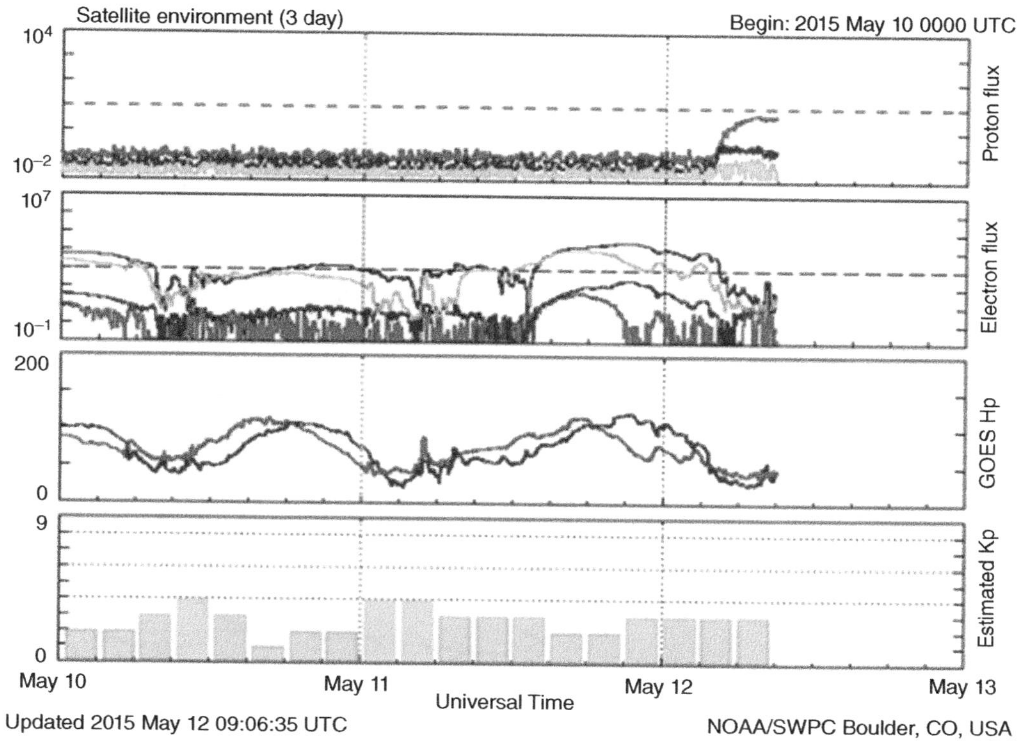

Figure 10.2 Rise in proton flux before the heat wave in May 2015.

Total electron content on the specific location of the Earth has shown sympathetic relation with the proton flux measured by SOHO satellite (Figure 10.4).

Shift of the Earth's poles is controlled by the changing Sun. The axis of rotation of the Earth and its relative gyroscopic adjustment further influence the total electron content (Figure 10.5).

Global Extreme Heat wave of 2021 during the Pandemic of 2019–2021 has limited man made activities. It is possible that the rise in temperature is due to extraterrestrial changes. The Sun is showing occasional rise in Proton flux which is trapped in the Van Allen belts.

This was just before the anomalous rise in atmospheric temperature in the Himalayan region, which was initiated by the release of heat energy from the trapped proton drift. "After this event, anomalous rise in cosmic ray was recorded. Rise in cosmic rays was instrumental in condensation of the clouds leading to the cloudburst in Kedarnath."

It took 20 days and 6 hours for the mechanism (of heat transfer to cloud appearance) to initiate the cloudburst in Kedarnath. The heat from the Sun was captured in the Van Allen belt (between the Sun and Earth), which further accelerated the protons. This extraterrestrial influence led to rise in temperature to release the aerosol trapped in the glaciers and atmosphere in the Indo-China border, initiating the cloudburst (Figure 10.6).

Generally, increase in aerosol heating over the Indo-Gangetic plains in the pre-monsoon period leads to a strengthening of the Indian monsoon. The heat transfer from the protons to the atmosphere has affected not only the atmospheric water vapor but has been responsible for melting of glaciers that feed the river Ganges.

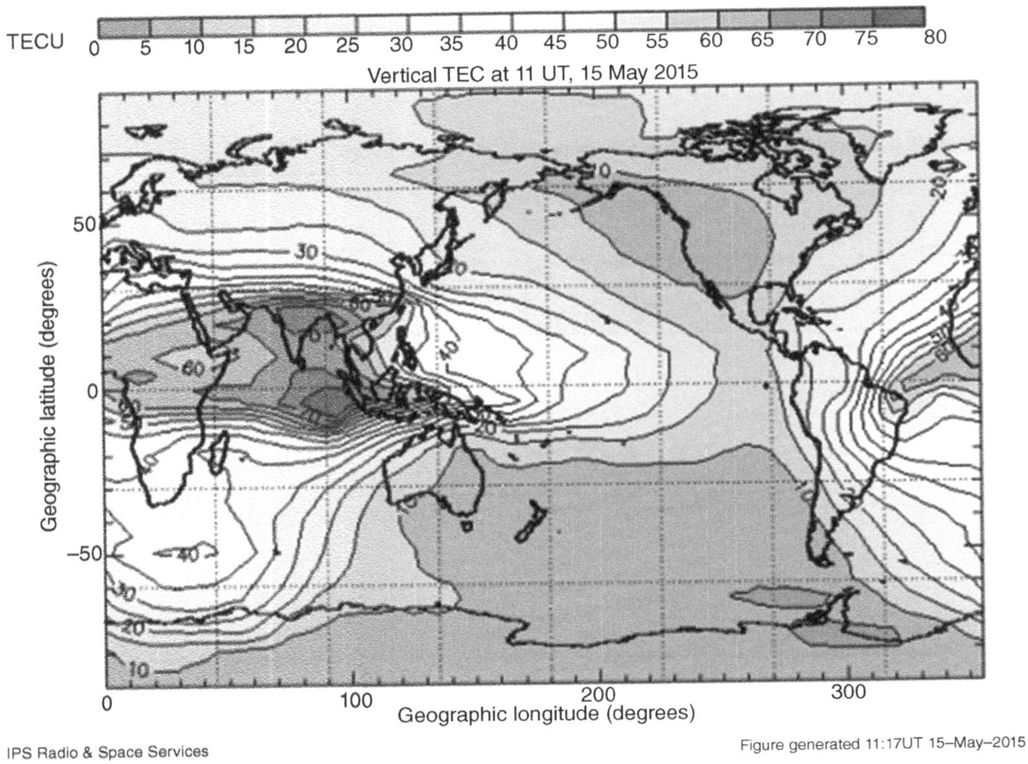

Figure 10.3 Total electron content rises in India before heat wave. Source: Adapted from https://www.sws.bom.gov.au/Satellite/2/2.

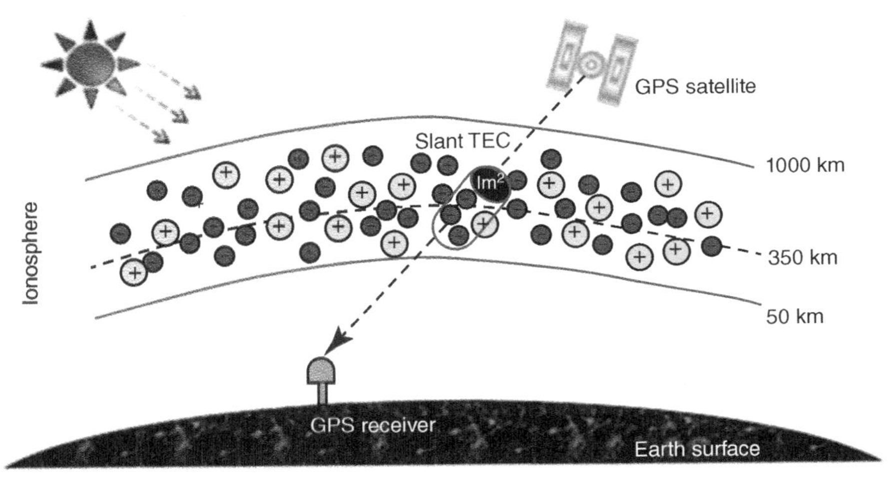

Figure 10.4 Schematic diagram of mechanism of total electron content by GPS receiver. Source: Adapted from N. Ya'acob et al. [1].

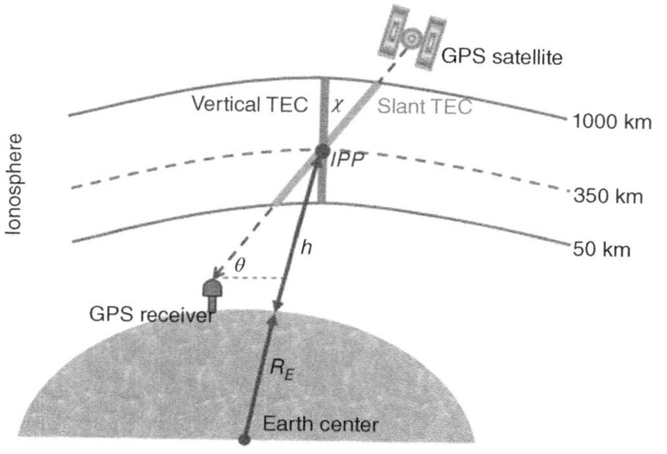

Figure 10.5 Showing the changing Earth and subsequent total electron content. Source: Adapted from N. Ya'acob et al. [1].

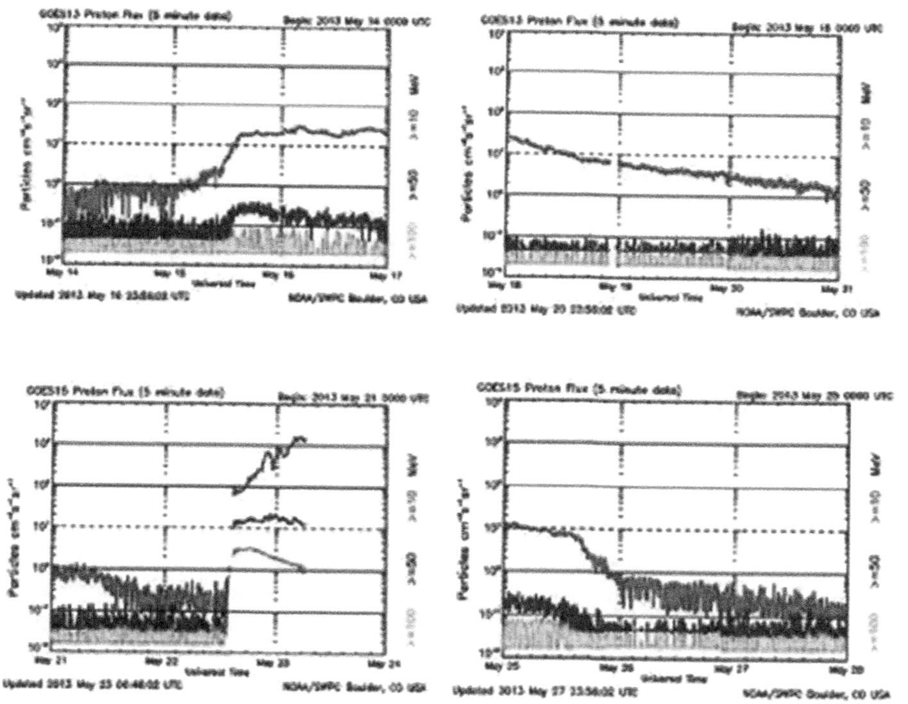

Figure 10.6 Rise in proton flux from 15 to 26 May to accumulate heat in the geomagnetic field line.

The influence of the Sun, along with anthropogenic activities, on climate change needs more study, especially its manifestation in torrential rains. "This is a radical departure from previous principles but is consistent with existing observations." This hypothesis can be established as Mukherjee extraterrestrial correlation if it repeats further. The new concept does not change the general conclusion that increased proton flux from the Sun reserves trapped heat in geo-specific locations, which influence temporary changes in the atmosphere (Figure 10.7).

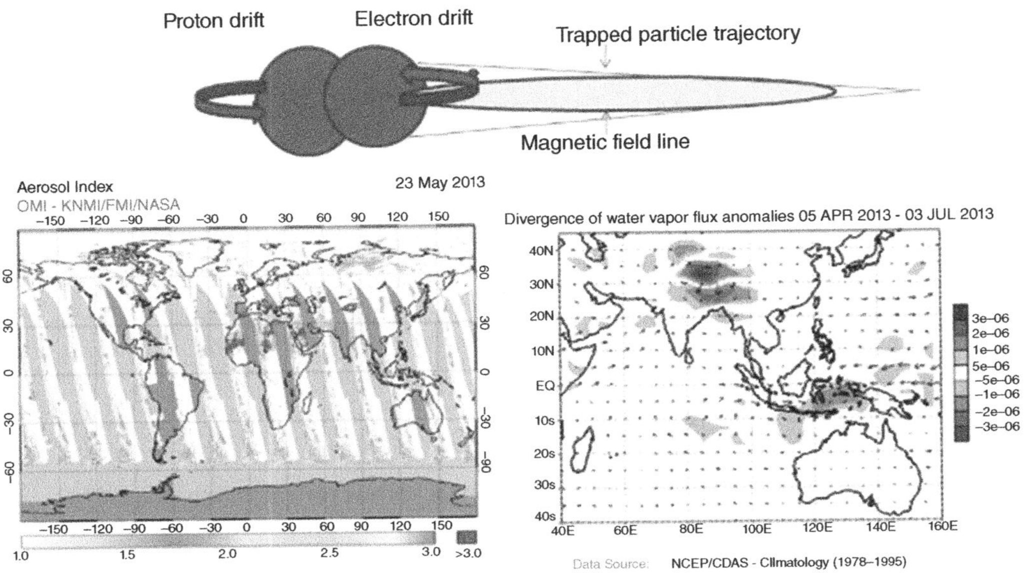

Figure 10.7 Mechanism of Cloudburst in Kedarnath, Uttarakhand, India. Charged particle motion in the magnetosphere initiates heating of glaciers and atmosphere in Northern India and China to release trapped Aerosol, which creates water vapor anomaly from 5 April to 3 July 2013.

The Kedarnath extreme weather event is a clear manifestation of climate change due to sudden global warming.

The devastating flood in Uttarakhand, India, in mid-June 2013 was a combined impact of cloudburst in Uttarakhand, quick melting of glacier at high altitude due to beating of ice sheet by raindrops and breaching of natural embankment of Chorabari Tal (north of Kedarnath) due to accumulation of excess surface runoff (Figure 10.8). Within 48 hours, 280 mm rainfall occurred and about 5 ft of snow was precipitated at higher altitudes. Apart from Uttarakhand, Himachal Pradesh, northeast Rajasthan, and New Delhi also received torrential rainfall. For New Delhi, it was an advent of early monsoon that broke the past record of 150 years. In western India, heavy rainfall occurred during mid-June in Gujarat and southern Rajasthan due to early advent of Arabian Sea branch of monsoon. The monsoon winds were pushed northeastward by an upper air current. On its way northeast, the cloud gathered moisture as a result of evaporation from freshwater bodies and irrigated agricultural fields in its path. Climate change manifestation can be seen in the recent disaster of Kedarnath, Himalayas, India. Influence of the Sun combined with the anthropogenic activities may be responsible for the catastrophe. Steep rise in solar proton flux above 10 MeV for 12 days from 15 to 26 May 2013 has been recorded by SOHO satellite. During the same period, the cosmic ray intensity was recorded at an all-time high in Jawaharlal Nehru University (which is representing regional cosmic ray data in SEVAN of Asian office of Aerospace Research). Prior to the cloudburst in Kedarnath area of Uttarakhand, Himalayas, the abnormal rise of the atmospheric temperature in this area was initiated by the release of heat energy from the trapped proton drift in the magnetic field line. The proton flux has the potential to be trapped for a long time in the geomagnetic field, enhancing the ionization process and heating of the upper part of the atmosphere in the proximity of the Uttarakhand area. After this event, an anomalous rise in cosmic ray was recorded. Changes in ionization affect the abundance of aerosols that serve as the nuclei of condensation for cloud formation. The rise in cosmic rays were instrumental in

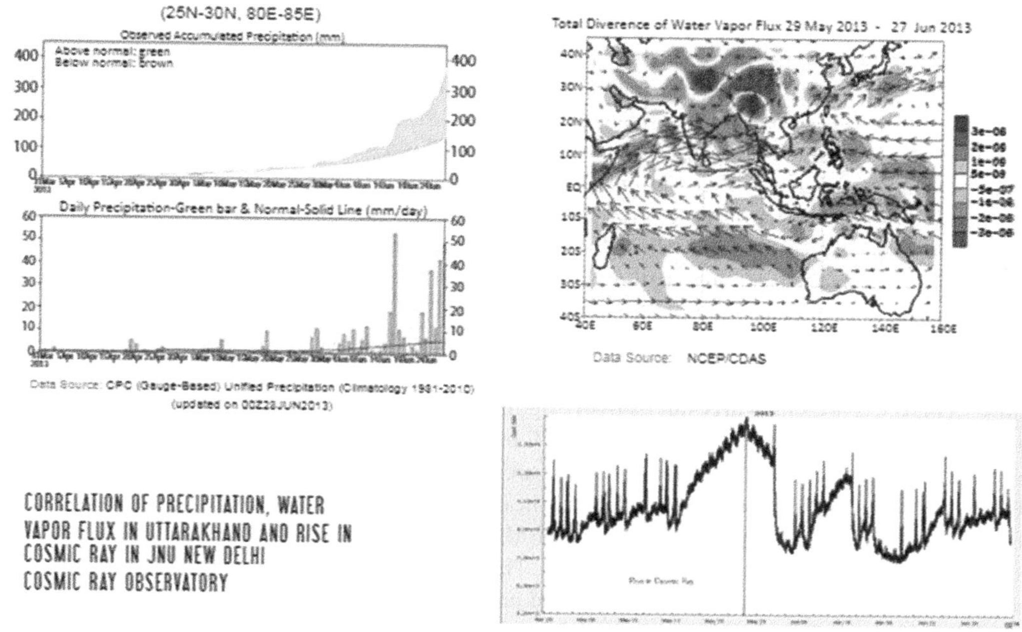

Figure 10.8 Sudden downpour of rainfall in Himalayan region is due to rise in evaporation, due to persistent heat and galactic cosmic rays.

condensation of the cloud leading to the cloudburst in Kedarnath. Improper land use change in this area along with the extraterrestrial influence led to a rise in temperature, releasing the aerosol trapped in the glaciers and atmosphere at the Indo-China border to initiate the cloudburst.

Ejection of toxic gases into the atmosphere from industrial plants, transportation, individual fireplaces, etc., has been the problem that modern humanity must resolve as soon as possible. Air pollution is reflected in both living organisms and the thermal properties of the atmosphere, especially in terms of anticyclonic weather conditions. However, the subject of scientific debates for many years is to what extent the temperature changes are the result of the natural processes and how much man by his actions causes these changes. In this study, we briefly introduce the newest standpoints of the IPCC, as well as the results of the research of the scientists who are classified into the so-called opposition science.

Regardless of the rapid development of science and technology at the present time, theoretical considerations have not yet reached such a level that would uniquely explain the causes of the formation of the long-period fluctuations and trends of climate fluctuations in the direction of cooling or warming. Even more serious has been the prediction based on quantitative data of future climate change. To provide energy assessment of possible causes of climate changes, one must start from the equilibrium balance of the thermodynamic energy, from which it follows that the mean temperature of the atmosphere of the planet is set by the following parameters:

- The size of the received solar radiation
- The ability of reflecting solar radiation, that is, albedo
- The losses occurred by long-wave radiation of the Earth's surface
- Turbulent heat exchange with the surface.

Part of the energy is converted into kinetic energy by the motion of the atmosphere. In addition, the energy circulation of the atmosphere leads to a redistribution of heat energy on the planet. The complex interactions of all these parameters lead to what we observe as fluctuation and climate changes at different time intervals.

The Intergovernmental Panel on Climate Change (IPCC) is a scientific intergovernmental body under the auspices of the United Nations. It was established in 1988 by the World Meteorological Organization (WMO) and the United Nations Environment Programme (UNEP). The IPCC produces reports that support the United Nations Framework Convention on Climate Change (UNFCCC), which is the main international treaty on climate change. IPCC reports also contain a "Summary for Policymakers," the concise and most cited IPCC documents.

The ultimate objective of the UNFCCC is "to stabilize greenhouse gas concentrations in the atmosphere at a level that would prevent dangerous anthropogenic interference with the climate system". The IPCC does not carry out its own original research nor does it do the work of monitoring climate or related phenomena itself. The IPCC bases its assessment on the published literature, which includes predominantly peer-reviewed sources.

Taking on the role of institution in awakening the conscience of mankind, the IPCC in the reports periodically and publicly expressed its views on the possible future climate changes caused by, first of all, increase in the concentration of CO_2. Thanks to its work, the terms "global warming" and "greenhouse effect" have become a standard part of scientific vocabulary. In IPCC report the emphasis is given on the rise in temperature. On the other hand, there are quite a few authors who consider the IPCC projections oversized, citing strong arguments and scientifically verifiable facts. This seems to be the case also with studies on the increase in atmospheric carbon dioxide content. Here, many studies have been performed, but many of them are not found by our search string that is concentrated on climatic consequences of increased levels of greenhouse gases and not on the atmospheric chemistry in its background. In the example of "Summary for Policymakers" from 2001, it can be seen that the standpoint of the IPCC is that anthropogenic influences are dominant in the contemporary climate variations, although in the report itself there are inconsistencies and ambiguities.

What can be inferred about it; what is the impact of human activities on fluctuation and climate change in the twentieth century? It stands in the "Summary for Policymakers" that it is unlikely that the warming in the last 100 years is solely the consequence of the self-regulation of the climate system. Also, the reconstructed data for the last 1,000 years indicate that it is unlikely that it is caused entirely by natural causes. However, at the end of the same paragraph, it is said: "Nevertheless, the observed warming in the first half of the twentieth century could be attributed to the natural factors." The sentence from the previous paragraph merges with the last sentence where it is stated that "there are new and compelling evidences that most of the warming registered in the last 50 years is attributable to human activities."

Talking about the rise in air temperature in the twentieth century in the first paragraph, it is said that "the observations showed a high degree of variability, and that most of the warming during the twentieth century occurred during two periods: 1910–1945 and 1976–2000." This means that possible anthropogenic greenhouse effect could dominate only in the last quarter of the twentieth century, as the natural factors are the cause of the temperature change in the first half of that century, and significant change in temperature is not observed in the period 1946–1975.

The increase in the average temperature on the surface of the planet in the twentieth century was $0.6 \pm 0.2\,°C$. However, as noted above, this increase was predominantly carried out in two periods: 1910–1945, when there was no anthropogenic impact, and 1976–2000, when there was anthropogenic impact according to the IPCC. This means that the growth was taking place not only in the last quarter of the twentieth century, when it could possibly be dominated by anthropogenic greenhouse effect, but part of this increase can undoubtedly be attributed to natural factors, up to 1945. Nevertheless, the impact of the development of urban heat island on the rise is possible. From all of these, it is clearly implied that the increase in planetary temperature of 0.6 ± 0.2 °C cannot be solely attributed to anthropogenic impact.

In a document of the Report of Working Group I of the Intergovernmental Panel on Climate Change, Summary for Policymakers, from 2013, is the following: "It is extremely likely that more than half of the observed increase in global average surface temperature from 1951 to 2010 was caused by the anthropogenic increase in greenhouse gas concentrations and other anthropogenic forcings together. The best estimate of the human-induced contribution to warming is similar to the observed warming over this period. Greenhouse gases contributed a global mean surface warming likely to be in the range of 0.5 °C to 1.3 °C over the period 1951 to 2010, with the contributions from other anthropogenic forcings, including the cooling effect of aerosols, likely to be in the range of −0.6 °C to 0.1 °C. The contribution from natural forcings is likely to be in the range of −0.1 °C to 0.1 °C, and from natural internal variability is likely to be in the range of −0.1 °C to 0.1 °C. Together these assessed contributions are consistent with the observed warming of approximately 0.6 °C to 0.7 °C over this period."

Therefore, the contribution to global warming from natural forcing and from natural internal variability is in the range between −0.2 and 0.2 °C over the period 1951–2010. It is not sure if the natural factors influenced the decrease or increase in temperature, or they had any impact, either individually or in total. The range of estimate of the impact of anthropogenic factors is significantly higher, whereas a sign for the effect of aerosols is not certain. If we take the lower limit of the contribution for greenhouse gases of 0.5 °C and the upper limit for aerosols (−0.6 °C), it turns out that the impact of aerosols on temperature was higher. Of course, this combination of impact does not fit the estimated increased value of 0.6–0.7 °C over this period. However, it is clear that anthropogenic factors dominate the increasing temperature and largely influence it by one-third (0.2–0.6 °C).

It is said in the same document that "In addition to robust multi-decadal warming, global mean surface temperature exhibits substantial decadal and interannual variability. Due to natural variability, trends based on short records are very sensitive to the beginning and end dates and do not in general reflect long-term climate trends. As one example, the rate of warming over the past 15 years (1998–2012; 0.05 [−0.05 to 0.15] °C per decade), which begins with a strong El Niño, is smaller than the rate calculated since 1951 (1951–2012; 0.12 [0.08 to 0.14] °C per decade)."

This would mean that in the period 1998–2012, there was no statistically significant increase in temperature. During the same period, the concentration of CO_2 has increased by 7%, which did not substantially have an impact on global temperature. It is also noted that the natural variability is present, which in this period is obviously of greater significance in relation to the anthropogenic factors. Although in the footnote it is said that "Trends for 15-year periods starting in 1995, 1996, and 1997 are 0.13 [0.02 to 0.24] °C per decade, 0.14 [0.03 to 0.24] °C per decade, and, 0.07 [−0.02 to 0.18] °C per decade, respectively"; however, a delay in the rise of temperature is not consistent with the models used by the IPCC. "However, for the 15-year trend interval corresponding to the latest observation period 1998–2012, only 2% of the 62 CMIP5 and less than 1% of the 189 CMIP3 trend computations are as low as or lower than the observed trend" (Figure 10.9).

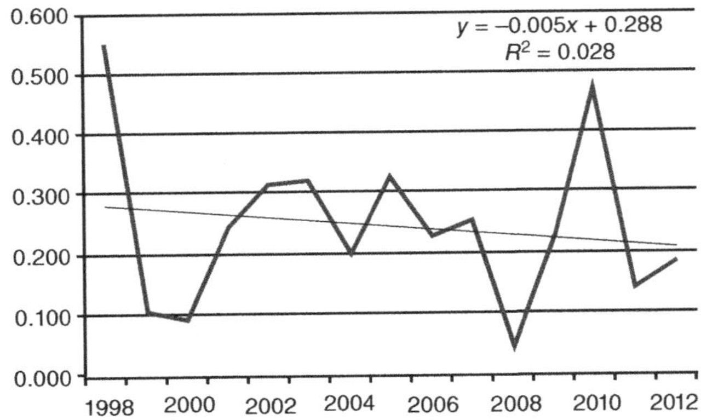

Figure 10.9 Remote sensing system (RSS) lowers troposphere anomaly temperature trend. Source: Adapted from https://www.remss.com/measurements/upper-air-temperature/.

Moreover, the set of satellite data remote sensing system (RSS) of temperature measurements in the first 8 km of the troposphere (latitude range 70.0 S to 82.5 N) showed a statistically insignificant cooling of −0.05 °C per decade (Figure 10.9). The RSS satellite dataset shows no global warming at all for 213 months from September 1996 to May 2014. That is more than half the entire 425-month satellite record. On the same satellite data, there has been no statistically significant global warming for more than 26 years. The results of a number of studies indicate that as a consequence of greenhouse forcing, general circulation models predict a greater positive temperature trend for the troposphere than the surface.

10.3 Arguments in Contradiction to the View of Global Warming

Bearing in mind the restriction of the scope of this study, we decided to give a brief representation of the results in the last four to five years. Analyzing the evolution of the scientific consensus on anthropogenic global warming in the peer-reviewed scientific literature, 11 944 climate abstracts from 1991 to 2011 were examined, matching the topics "global climate change" or "global warming." They found that 66.4% of abstracts expressed no position on anthropogenic global warming. In a second phase of this study, they invited authors to rate their own papers. Compared to abstract ratings, self-rated papers expressed no position on anthropogenic global warming in 35%. With reference to the papers, these authors concluded that the narrative presented by some dissenters is that the scientific consensus is on the point of collapse, while the number of scientific heretics is growing with each passing year. Contrary to this analysis, Lockwood argued that the popular idea (at least on the Internet and in some parts of the media) that solar changes are some kind of alternative to GHG forcing in explaining the rise in surface temperatures has no credibility with almost all climate scientists.

The impression is that during the last few years, more and more articles appear in which attempts are made to explain the current temperature changes by the Sun–Earth connection; they have recently re-evaluated the derivation of open solar flux and mean solar wind speed from geomagnetic activity data, using a better correction for the processing of the interplanetary field by longitudinal solar wind structure between the surface of the Sun and the Earth. The results are very similar to some workers in past, who used an averaging procedure to effectively make the same correction. The implications of the derived changes in open solar flux have been reviewed by several authors. Those authors argue that the current low solar cycle minimum is part of the fall from the grand solar maximum that has persisted during the space age. From the linear extrapolations (over short intervals), these authors predicted that the Sun will fall out of the maximum (defined by the mean level exceeded in 1920) within the interval 2011–2027. This agrees with the independent predictions for a consistent threshold, based on the distribution of durations of past grand solar maxima in cosmogenic isotope data.

A significant number of studies show that some meteorological parameters, such as storms and droughts, show a good correlation with the 22-year Hale magnetic cycle. The solar activity is the 11-year cycle, and it is this cycle that is mostly used for correlative analyses with meteorological parameters claimed that models predict that perturbations can descend from the stratosphere to the surface by altering the propagation of planetary waves, an effect that has been reported in observations. Explaining this "top-down" model, these authors suggest that the solar effect on temperature in the lower equatorial stratosphere can influence the jet streams in the underlying troposphere. Propagation of the charged particles to the surface is also studied in many other papers, with that conclusion that the influence of environment on these processes could be much more intensive than it was previously.

Quite contrary to the IPCC prognosis, results of this research indicate an annual average temperature drop of 0.9 °C in the Northern Hemisphere during solar cycle 24. For the measuring stations south of 75N, the temperature decline is of the order 1.0–1.8 °C and may already have started. For Svalbard a temperature decline of 3.5 °C is forecasted in solar cycle 24 for the yearly average temperature. An even bigger temperature drop is forecasted in the winter months. An Arctic cooling may relate to a global cooling in the same way, resulting in a smaller global cooling, about 0.3–0.5 °C in SC 24. From correlation studies of 7 (not all global) temperature series for the period 1610–1970, a solar contribution of 41% to the secular temperature increase is found. Analyzing global temperature curves for periodic oscillations, Scafetta concludes that the climate is forced by astronomical oscillations related to the Sun, and at least 60% of the warming since 1970 can be related to astronomical oscillations. According to Scafetta, our planet will experience Little Ice Age with maximum cooling around the middle of this century.

The indications that the solar–climatic coupling is nonlinear were already obtained in several works. Specifically, this can be confirmed by the experimental fact that the temperature response of the atmosphere to the solar radiation variability during the Schwabe cycle is a factor of 2–3 as large as such are responses following from thermodynamic considerations. Numerous evidences of the effect of solar cosmic ray fluxes, related to powerful flares, on the processes in the Earth's lower ionosphere were obtained in many works. It is interesting to note that the quasi-thirty-year climatic periodicity is often observed in different climatic processes and is called the Brückner periodicity. If the climatic system is strongly nonlinear, even low-energy SA manifestations in the Earth's lower troposphere can actually cause substantial variations in the global climatic system.

Reference

1 Ya'acob, N., Abdullah, M., and Ismail, M. (2010). GPS Total Electron Content (TEC) prediction at ionosphere layer over the equatorial region. INTECH open access publishers. ISBN 9533070722.

Further Reading

Davies, C.J. and Constable, C.G. (2020). Rapid geomagnetic changes inferred from Earth observations and numerical simulations. *Nature Communication* 11: 3371.

Mukherjee, S. (2014). Extra terrestrial remote sensing and geophysical applications to understand cloudburst in Uttarakhand India. *Journal of Geophysics and Remote sensing* 3 (3): 1000124.

Mukherjee, S. (2021). Heliosphere-Solar-Terrestrial Impact on Himalaya India. *XVIIth HvAr Astrophysical Colloquium, The Sun and Heliosphere* (20–24 September 2021) [Online]. HvAr observatory Faculty of Geodesy, University of Zagreb, Croatia. https://oh.geof.unizg.hr/images/meetings/XVII-HAC/xvii-hac-abstract-book.pdf

11

Extraterrestrial Remote Sensing of Planets

11.1 Introduction

Terrestrial satellites create new datasets and make the life of the Earth dwellers comfortable every day. However, resources are limited and the rising population is forcing scientists to think and plan for the extraterrestrial remote sensing of planets. Besides mineral resources, water, and life-supporting oxygen, the thrust of extraterrestrial remote sensing is on finding life on the planets of solar system and in other planets of binary star systems across the universe. It is being investigated whether life may evolve due to geochemical processes on the levels of methane measured by the Cassini spacecraft on Saturn's icy moon. It is yet to be established that life exists on Enceladus, the largest moon of Saturn. It is approximately 500 km in diameter and about a tenth of Saturn's largest moon, Titan. Enceladus is mostly covered by fresh, clean ice, making it one of the highest reflective bodies of the Solar System. Existence of life on the moon of Saturn is under investigation; the results would be consistent with microbial activity similar to that known to occur at hydrothermal vents in Earth's oceans.

To investigate the planetary systems of the Sun and the moon of the planets, it is imperative to collect the information of their relative rotation and oscillation. All the planets have specific differential rotation than the Sun. However, all these systems have a common component that is the surface rotation, which is on certain latitudes quicker than that of a rigid body. This phenomenon shows a self driven motor must have been instrumental for these rotations. All differential rotation is parallel or almost parallel to the rigid body, which behaves uniquely and differently in planetary bodies.

The atmosphere of the planets or precisely the gaseous envelope also rotates although its speed is much faster than the rotation of the specific planet. The relative rotation of planets, solid mass and gaseous body gives a clue to further investigate the possibility of a relationship between atmospheric pollution and stability of planets, including the Earth.

Science on terrestrial tectonics so far neglects the rotation of the Earth. It is based rather on paleontology, forms of coast, and magnetic materials. There is a relationship between rotation of the planets and its magnetic field (Figure 11.1).

Cassini-Huygens sensors investigated the salty plumes belching from the interior of Saturn's moon Enceladus. The findings were exciting; the satellite Cassini revealed some new findings. Enceladus was extensively investigated by the Cassini spacecraft, which reveals its surface and environment in greater detail. Its sensors made an unexpected finding. The collected information by optical and microwave sensors were analyzed remotely. It reveals that compounds present there are of similar geological materials to that of the Earth. The presence of rocks like those on Earth is also associated with hydrothermal vents on Earth's ocean floor, which further supports that these planets were formed together at the same time.

Extraterrestrial Remote Sensing and Climate Change, First Edition. Saumitra Mukherjee.

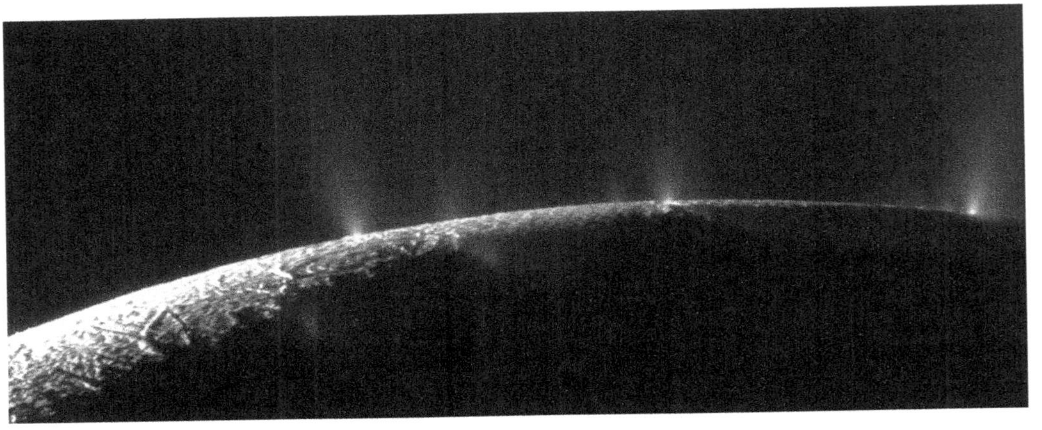

Figure 11.1 Methane detected on Enceladus, a moon of Saturn, shows signs of life. Source: NASA.

It remained possible that known changes in thermodynamic equilibrium is responsible for geochemical evolution, which is a non-biological process that could be responsible for the abundance.

The process of methane formation in Enceladus is yet to be established. Decomposition of foraminifers in the deep sea of the Earth has been recorded to be instrumental in methane formation. It has not been established how the amount of methane observed spewing from Enceladus is occurring. That means it could be an unknown process – or it *could* be biological in origin.

Like microbes in the Earth's ocean, there may be microbes on Enceladus that "eat" the dihydrogen and produce methane, explaining the surprisingly large amount of methane detected by Cassini. Searching for such microbes, known as methanogens, at Enceladus' seafloor would require extremely challenging deep-dive missions.

Mathematical modeling of the remotely sensed data appears to be using known variables – processes that produce methane here on Earth. Hydrothermal vents, which Cassini sensors detected in Enceladus' plumes, included, along with methane, dihydrogen and carbon dioxide. The hypothesis based on remotely sensed data supported by bio-geochemical shows the possibility of relative abundances of these hydrocarbon producing compounds.

The investigation has sufficient proof to confirm the dihydrogen abundance and determine if it could be produced by hydrothermal activity. The next step was to determine if it was sufficient to feed a population of hydrogenotrophic methanogens. In correlation with the spectral response on the Earth, these are Archaea (single-celled microorganisms) that metabolize molecular hydrogen and carbon dioxide to produce methane (Figure 11.2).

Exploration of another important planet of the solar system has been initiated by spacecrafts Pioneer 10 and Jovian system in 1973 by NASA. Extraterrestrial remote sensing established magnetic field in Jupiter. However, its intensity is not the same as that of the Earth. Planetary magnetic fields were formed by strong convective currents in the core of the planets. These currents vary depending on the speed of rotation of different layers of planetary system.

11.2 Radio Emissions of Jupiter Were Inferred to Establish Its Magnetic Field

Magnetic field highlights from Planetary Mission (Table 11.1) shows the variation of magnetic fields in different planetary bodies.

Table 11.1 Variation of magnetic fields in different planetary bodies.

Planet	Spacecraft	Date	Magnetic highlights
Mercury	Mariner 10	1974–1975	• Detection of dynamo-generated magnetic field • Field characterized by weak dipole moment, large dipole offset, and small dipole tilt
	MESSENGER	2008–2015	
Venus	Mariners 2, 5, 10	1962, 1967, 1974	• No dynamo-generated magnetic field detected • Detection of small induced magnetosphere due to solar wind–ionosphere interaction
	Venera 4, 9, 10	1967, 1975, 1975	
	Pioneer Venus Orbiter	1979–1981	
	Magellan	1990–1994	
	Venus Express	2006–2014	
Earth	Magsat	1979–1980	• Mapping of dynamo-generated and crustal magnetic fields to high resolution • Detailed observations of magnetic field secular variation
	Orsted	1999–till date	
	Champ	2000–2010	
	SAC-c	2000–2004	
Moon	Lunar Prospector	1998–1999	• No dynamo-generated magnetic field detected • Detection of localized crustal magnetic fields
Mars	Mars Global Surveyor	1997–2007	• Detection of strong localized crustal magnetic fields
Jupiter	Pioneer 10, 11	1973, 1974	• Detection of dynamo-generated magnetic field, similar in morphology to Earth's field
	Voyager I, II	1979	
	Galileo	1995–2003	
Galilean moons	Galileo	1995–2003	• Detection of Ganymede's dynamo-generated magnetic field • Detection of induced magnetic fields in global subsurface oceans of Europa, Ganymede, and Callisto
Saturn	Pioneer 11	1979	• Detection of dynamo-generated magnetic field • No non-axisymmetric field components detected, upper limit on dipole tilt: 0.06°
	Voyager I, II	1980, 1981	
	Cassini	2004–2017	

Table 11.1 (Continued)

Planet	Spacecraft	Date	Magnetic highlights
Titan	Cassini	2004–2017	• No dynamo-generated magnetic field detected
Uranus	Voyager II	1986	• Detection of dynamo-generated magnetic fields • Field characterized by significant non-dipolar, non-axisymmetric components
Neptune		1989	

Source: Adapted from Doris Breuer [1].

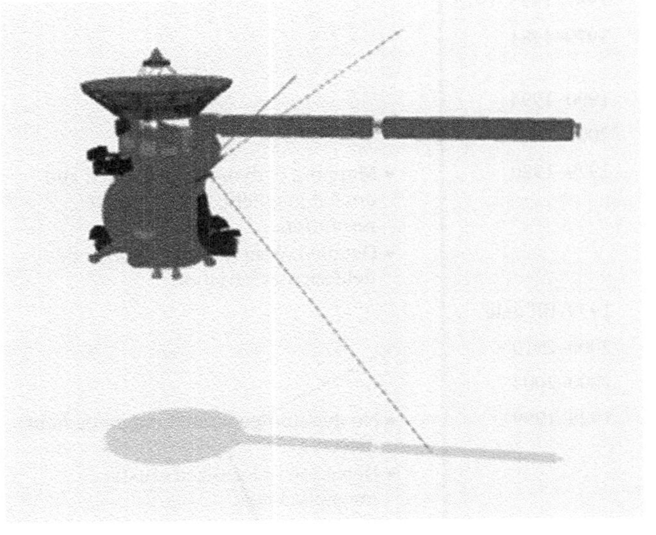

Figure 11.2 Cassini spacecraft of European Space Agency, dedicated to probe Saturn and its moons from 2005 to 2017.

The comparative table elaborates that magnetic field of Jupiter is comparable to that of Earth; however, the atmosphere of Jupiter is much different in comparison with the Earth's atmosphere (Figure 11.3).

The information of extraterrestrial remote sensing of Jupiter was further revealed by Voyager I and Voyager II probe. It was found that Jupiter has 79 moons; however, it has four that are the most prominent, namely Io, Europa, Ganymede, and Callisto (Figure 11.4).

Extraterrestrial satellite Ulysses further studied Jupiter's magnetosphere in 1992 and 2000. The Cassini probe investigated remotely sensed data by Cassini and new Horizon images to identify ice crystals and atmosphere in Jupiter as well as its moons. Other European member countries, China, and India are likely to launch new satellites to study the extraterrestrial remote sensing of Jupiter in the near future.

All gas planets have more complicated differential rotation than the Sun. All these systems have a common component that the surface rotation is quicker on certain latitudes than that of the

Figure 11.3 Jupiter exploration program has been conducted by automated spacecraft known as Pioneer 10 and Jovian system. Source: NASA/Wikimedia commons/Public domain.

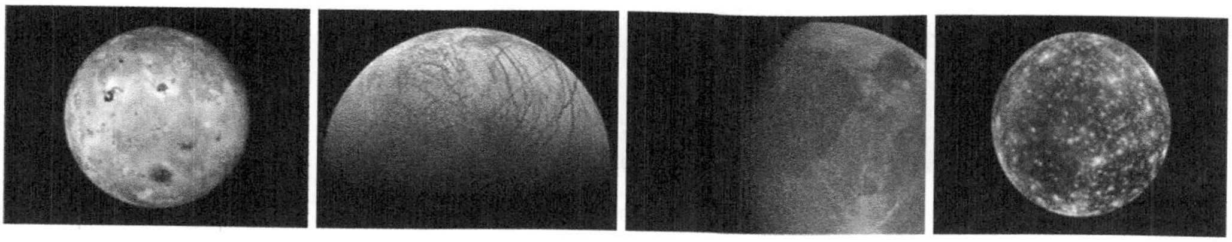

Figure 11.4 Four prominent moons of Jupiter: Io, Europa, Ganymede, and Callisto. Source: NASA/Wikimedia commons/Public domain.

rigid body. This phenomena further proves the possibilities of a self driven motor for the rotation. All differential rotations are parallel to almost parallel to the rigid body. Therefore, all these differential rotations seem to prove the gyroscope model. Two components are important here, orientation and angular velocity. Density and gravity of the solid mass or rocks of planetary bodies and gaseous material above are being inferred by using gravity recovery sensors.

Venus rotates very slowly in 243 days to complete one rotation, however its atmosphere rotates much faster and it is 50 to 60 times quicker to take one rotation in only 4.2 days. The relative high gas velocity exists only at an altitude of 50 km; it is approximately zero on the solid surface. The zero velocity on the surface is clear. Whirls roll on the surface of Venus as the wheels of an auto roll on the road. The model of the granules is the most important idea of the solar gyroscope model. The Venus probe records data that justifies this model. The high velocity of the wind at the altitude of 50 km is also proof of the rolling whirls with wind speed of 140–160 m s^{-1} (Figure 11.5).

11.3 Planets Influenced by Pleiades

Pleiades is a constellation of stars in proximity to the Milky Way. Pleiades is also known as Seven sisters and Messier 45, which is an open star cluster containing middle aged, hot B-type stars in the northwest of the constellation Taurus. The relative position of the constellation of stars shifts in the magnetic field alignment of specific planets in a time-series shifting process. From 31 March 2012 to 2 April 2012, it was recorded that the Pleiades constellation came closer to the Venus of our planetary system (Figure 11.6).

The action begins at sunset on Monday, April 2, when Venus enters near the small bucket-shaped asterism. Sun facing part of the Venus can be seen by binoculars towards western end far away from the Sun. The combination can be observed immediately. The best time to see it is during the night

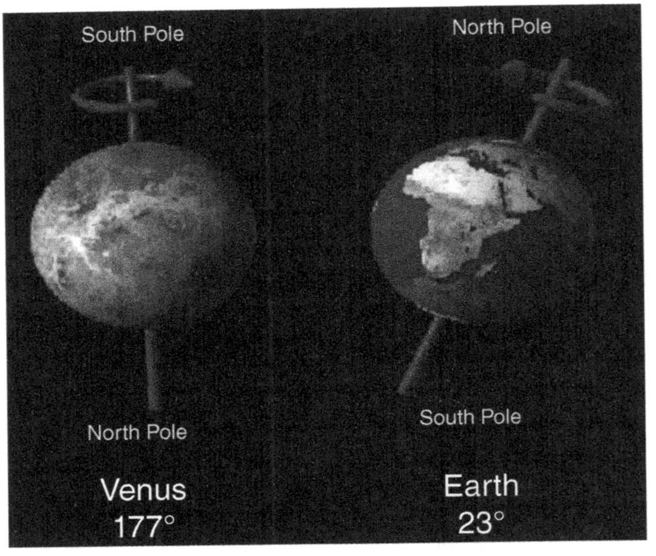

Figure 11.5 Tilting axis variation of Venus and the Earth. Source: NASA/JP.

Figure 11.6 Venus is closer to the Pleiades, 31 March 2012, and is photographed by astronomy professor Jimmy Westlake of Stagecoach, Colorado. Source: NASA/Wikimedia commons/Public domain.

of Tuesday, April 3, when bright Venus position in the south of the Sun-Earth Venus bucket. Venus will be on the handle of the bucket, on Wednesday, April 4. The planet passes through Pleiades, thus approximately once every eight years (Figure 11.7).

On 31 March 2012, there was a very powerful bolt of lightning in the vicinity of Woodward, Oklahoma, that spawned these red sprites. Sprites are electrical discharges that come out of the top of thunderclouds, unlike ordinary lightning bolts that plunge toward Earth. Sprites can tower as high as 90 km above ground. That makes them a form of space weather as they overlap the zone of auroras, meteors, and noctilucent clouds. Because they are associated with lightning, sprites are most often seen in summer months; however, sometimes they are visible in lower latitudes also during solar minimum when a specific constellation of stars like Pleiades comes effectively close to the planetary system.

Figure 11.7 Influence of the Pleiades constellation on the sky of Woodward, Oklahoma, shown as sprite at a height of approximately 90 km above the Earth. Source: Yuri Beletsky.

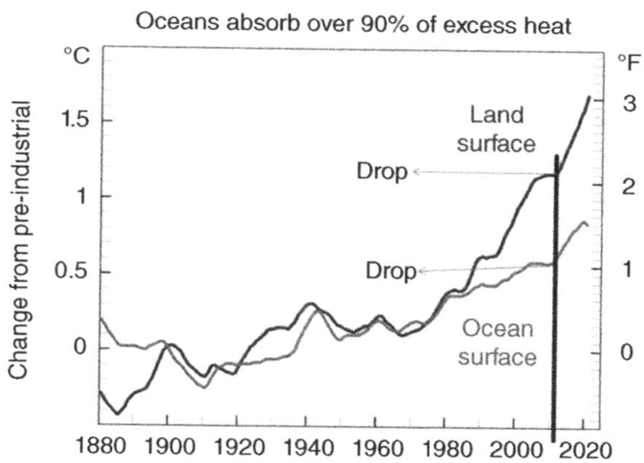

Figure 11.8 Appearance of sprite during drop in sunspots on 31 March 2012. Source: Adapted from Lina Tran 2020.

It has been observed that between 31 March and 2 April 2012, the atmospheric temperature and land temperature dropped due to reduced solar activity (Figure 11.8).

11.4 Influence of the Solar Storms Affects Saturn

Solar storms and thier effect on the planets are well established; however, magnetic field change has maximum impact among the variables of heliophysical parameters (Figure 11.9).

The effect of the solar storm on the Saturn environment has been inferred. It shows the transformation of the Sun before every solar maximum. When the industrial revolution began on the Earth 400 years ago, which were supposed to heat up the Earth, but in contrary to this the Earth faced a little Ice Age. Similar phenomena can be seen on other planets including Saturn. The impact of heliophysical parameters on the Saturn ring can be seen during solar minimum as an additional ice patch on it.

It has been recorded that the sudden decrease in coronal mass ejection after solar minimum has a similar effect on the extremely high snowfall in Greenland between 7 and 13 March 2011 with

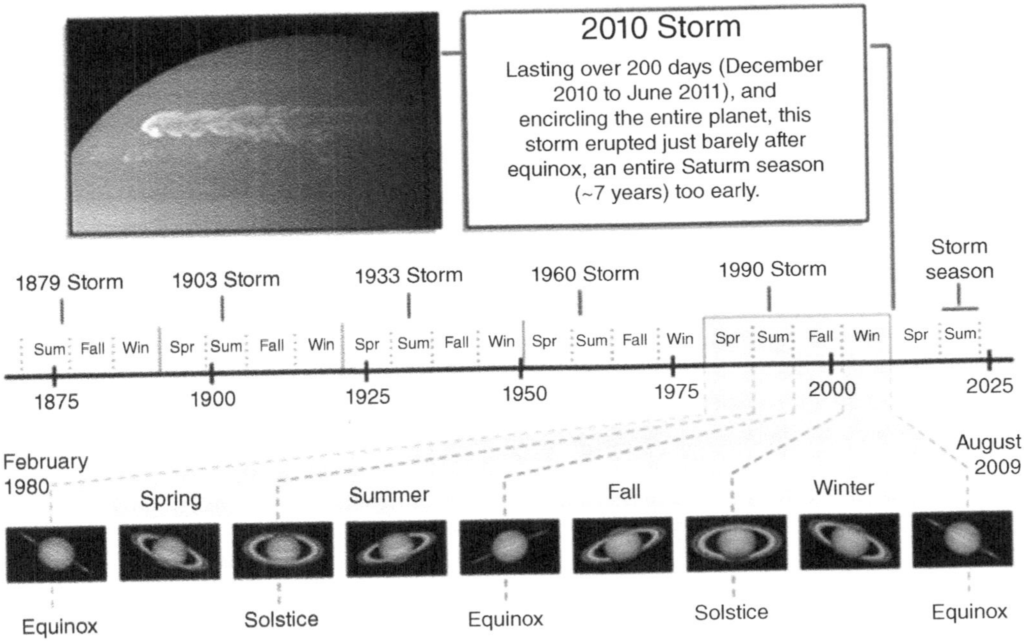

Figure 11.9 Effect of solar storm on Saturn environment from 1875 to 2025 (Projected).

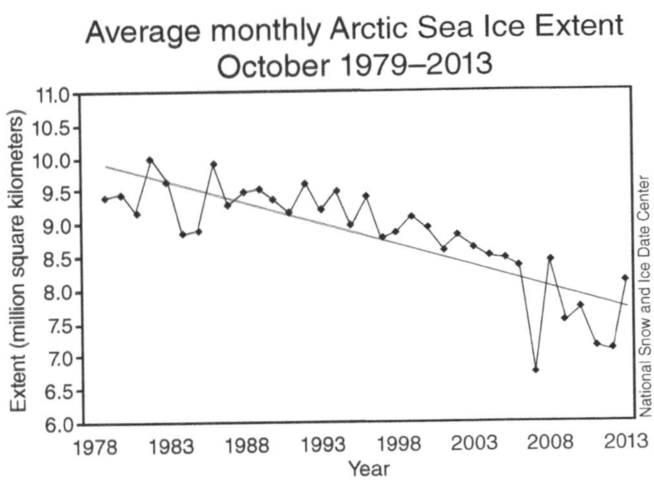

Figure 11.10 Rise in ice precipitation in 2013 during solar minimum. Source: Adapted from Chiodo et al. (2016).

subsequent rising of galactic cosmic rays (Figure 11.10). This phenomenon is known as Forbush decrease (Figure 11.11).

Different manifestations of effects on the Earth's weather systems were investigated in the world, which has been showing the influence of space weather on terrestrial changes, such as the 2005 Hurricane Katrina, Katia, etc. It is inferred that Radio Flux 10.7 has the greatest impact on wind speed of the hurricanes, except in the case of the Katia hurricane. In the general picture of the

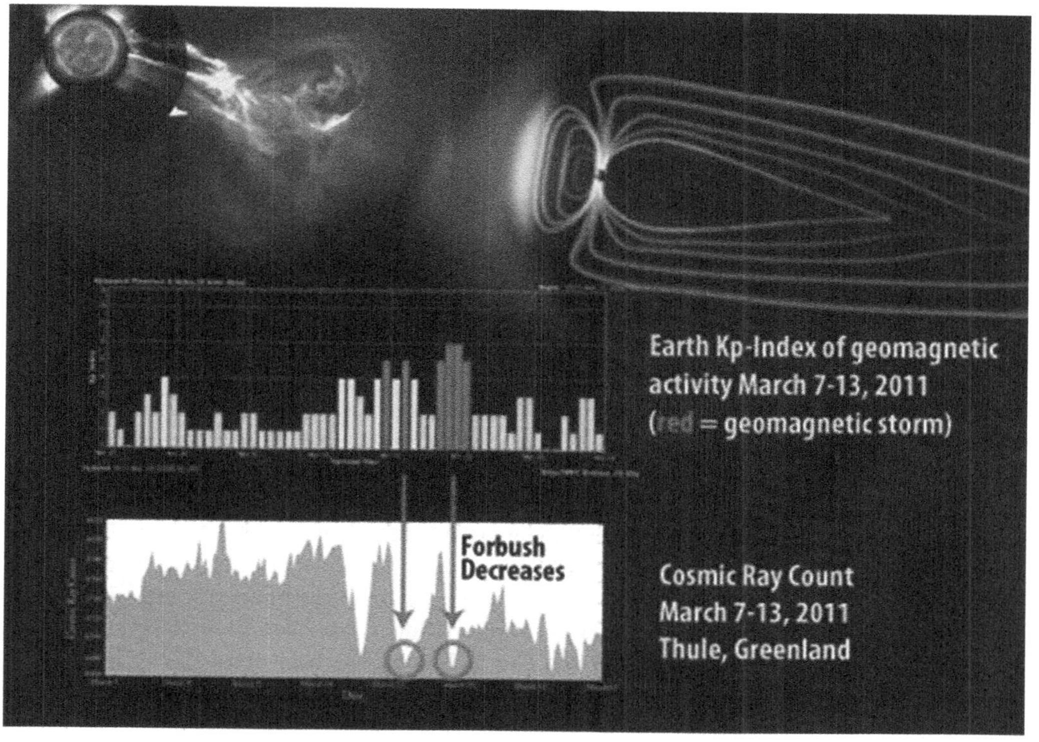

Figure 11.11 Lowering of galactic cosmic rays with rise in planetary indices (Kp) of the Sun.

change in pressure and wind speed over a longer period, there are other factors that were not always taken into account in the models.

References

1 Doris Breuer. (2021). Encyclopedia of the Solar System, Hardcover ASIN: B0182Q95YU.
2 Chiodo, G. et al. (2016). *Environ. Res. Lett.* 11 (3): 034015.

Further Reading

Affholder, A., Guyot, F., Sauterey, B. et al. (2021). Bayesian analysis of Enceladus's plume data to assess methanogenesis. *Nature Astronomy* https://doi.org/10.1038/s41550-021-01372-6.

Sun, J., Xu, T., Jiao, Z., and Wang, H. (2021). Mechanics of the gyroscopic precession and calculation of the galactic mass. *Preprints* 2021020206. https://doi.org/10.20944/preprints202102.0206.v1.

12

Remote Sensing of the Moon of the Earth

Recent extensional and contractional tectonic features present globally over the lunar surface have been studied to infer lunar crustal tectonism [1–4]. Spatial orientation of these tectonic features helps to elucidate the change in the interior geological dynamics of any planetary body with time [5, 6]. Investigation of indicators of recent crustal tectonics, such as fault lines, thrust fault scarps, and dislocation of debris along the identified fault planes, primarily using data from the miniature-synthetic aperture radar (Mini-SAR) aboard CHANDRAYAAN-1 mission, is the focus of this study. The ability of microwave sensors to penetrate the lunar regolith, along with application of m–χ decomposition method on Mini-SAR data, has been used to reveal unique features indicative of hidden tectonics [7–9]. The technique has been used to expose hidden lineaments and lobate scarps present within shadowed crater floors as well as over the illuminated regions of the lunar surface. The area around and within Cabeus B crater in the South Polar Region contains unique tectonic features indicative of relatively recent crustal movement [1, 2]. The study region should be focused more using a fine resolution and deeper penetrating radar sensor to study the uniqueness of the underlying geodynamics of the south polar lunar surface and derive comparisons with other polar and equatorial regions.

Crustal tectonism associated with crustal evolution deals with the perturbations taking place on the uppermost crustal layer of planetary bodies [10]. Differentiation of a planetary system into layers takes place to carry out the process of thermal evolution. This evolution is based on temperature-sensitive chemical differentiation causing the heat to drain out from the planet's interior similar to the conveyer belt system seen in the Earth's tectonic profile. The lunar crust thus formed is composed of lighter rock minerals underlying a thin layer of regolith. Further, as a planet cools down from the time of its formation, several geological changes take place both on the surface and on the subsurface due to changing stress strain regimes. The subsurface geological perturbations generated as a result often manifest in the form of various surface compressional and extensional deformations to accommodate the respective subsurface stress [11]. Surface features indicating extension form during interior warming whereas interior cooling develops compressional features [10]. Presence of crustal tectonic features on the lunar topography therefore acts as evidence of the state of the underlying geological activity. Besides, the relative orientation of the identified features collectively unveils the tectonic peculiarities of different regions on the lunar body.

The use of Mini-SAR data along with high-resolution Narrow Angle Camera (NAC) images from Lunar Reconnaissance Orbiter Camera (LROC) has been done to reveal the surface and subsurface features formed due to the associated compressive and extensive tectonic stress. Parameters such as circular polarization ratio (CPR) and the Poincare ellipticity factor (χ) derived from Stokes vectors (S_0, S_1, S_2, S_3) and degree of polarization (m) help in the estimation of the location and degree of

Extraterrestrial Remote Sensing and Climate Change, First Edition. Saumitra Mukherjee.

surface roughness enabling the identification of surface features formed as a consequence of the interior tectonic stress buildup and the subsequent accommodation of the associated strain by the crust. The features identified are otherwise not visible or faintly visible on optical satellite data as seen on the wide angle camera (WAC) images from the LROC.

"Lobate scarps" are surface traces of recent thrust faulting formed due to compressive stress acting on adjacent rock masses lying along the fault planes. They form due to movement along shallow, low-angle thrust faults with the hanging wall moving up relative to the footwall along the fault. They are considered to be the youngest of the landforms in geological time scales ($\leq$500 Myr), occurring as irregular segments connected together in a linear or curvilinear fashion, falling asymmetrically. In the present study, such scarps have been inferred within the South Polar Region of the moon using both optical NAC and Mini-SAR data derived $m-\chi$ decomposition images. The formation of such features may be associated with release of the compressive stress leading to decrease in the global lunar radius.

A distinct lobate scarp, traversing a length of approximately 20 km lying between Doerfel and Boltzman craters, inferred from NAC images, trends east–west between 71.29°S, 100.13°W, and 71.34°S, 103.01°W. It is segmented with a slight curvilinear pattern with the thrust sheet (hanging wall) partially cross-cutting medium- and small-sized craters (<500 m) along its length, which substantiates the recent formation of the feature (Figure 12.1d).

Alternately, compression marks can be seen on NAC images on the rock surface lying over the footwall, in a particular segment of the inferred scarp, possibly formed due to the compressive stress generated from the underlying convergence of rock masses along the thrust fault (Figure 12.2b). The formation of such marks could have occurred from the subsurface push primarily generated when the thrust sheet moves over the footwall. The compression marks suggest that this segment of the footwall has broken from the rest of the surrounding rock and therefore is experiencing compression both from the side of the fault plane as well as the broken end, leading to the formation of the compression folds.

It has been proposed that thermal stress may also have been responsible for the lineaments observed on the lunar lithosphere. Such lineaments are tectonic landforms, which indicate the occurrence of subsurface geological perturbations as a result of dislocation along fault planes resulting in surface disruptions, evidence of which gets buried over time under the lunar regolith or degraded due to space weathering. $m-\chi$ decomposition of Mini-SAR data has therefore been used to demarcate such hidden lineaments and associated dislocation features in the study region.

Other derived parameters from the SAR data such as the circular polarization ratio (CPR) further display the degree of surface roughness, which helps to demarcate any recent deformation along and around such inferred lineaments. CPR gives the ratio of the same sense backscatter power received as emitted over opposite sense backscatter power received as emitted. A higher surface roughness in the region of the deformation validates significantly recent formation of the feature since it has not been inundated or degraded from space weathering or by later nearby impact events [12]. To further understand the degree of surface roughness, the volume scattering contribution from $m-\chi$ decomposition was also observed. Higher CPR coincides with the region showing volume scattering. Hence, the $m-\chi$ decomposition images along with optical data elucidate the deformation features in greater detail to estimate their approximate time of formation in relation to other geological features located nearby.

Cabeus B, a relatively large impact crater centered around 82°18′S and 54°12′W, having a diameter of ~61 km, displays unique crustal activity. The gently sloping inner walls of the crater cover a broader area as compared to the inner walls of the surrounding craters (Figure 12.3a). The western boundary of Cabeus B cuts through half of another equally large crater. The optical images do not

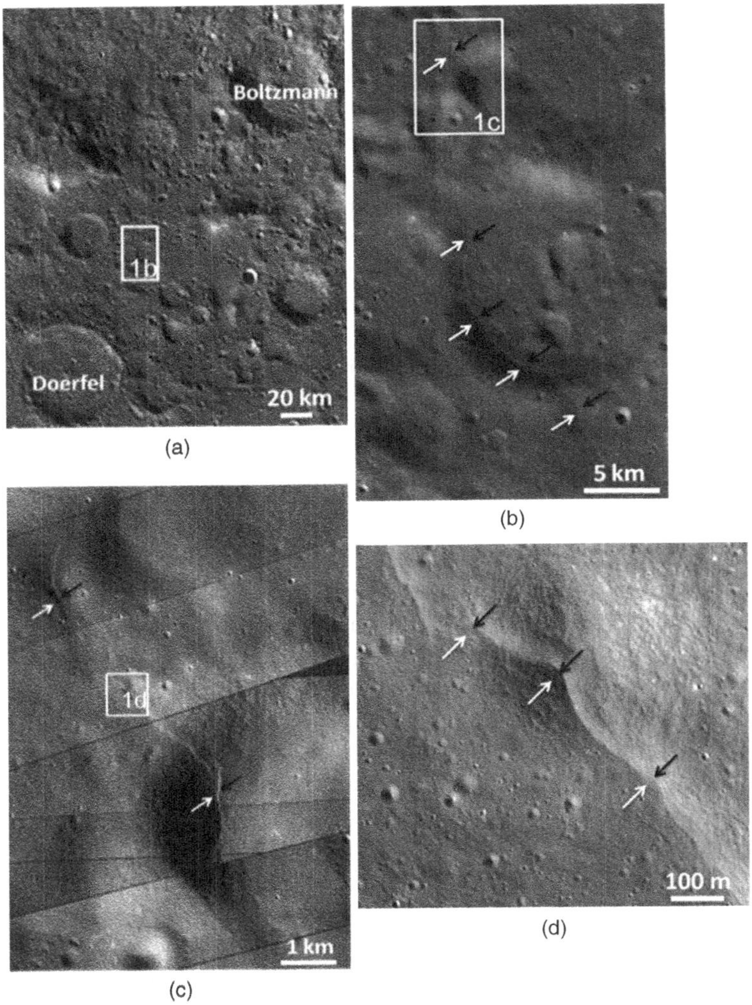

Figure 12.1 Scarp between Doerfel and Boltzmann craters (located between ~71.29°S, 100.13°W, and 71.34°S, 103.01°W). (a) Wide angle camera image of the inferred scarp lying between Boltzmann and Doerfel craters in (b). (b) Narrow angle camera image of the scarp region marked in (a) (white arrows pointing at the scarp). (c) Narrow angle camera image of the scarp region in (b) (white arrows mark the foot wall, black arrows mark hanging wall along the thrust fault). (d) Narrow angle camera image of region marked in (c) of a partially crosscut crater with ~200 m diameter.

show any fault lines present in this region. However, $m-\chi$ decomposition image reveals a lineament, ~62 km in length, cutting across the northern crater wall. Since the lineament partially lies in the shadowed part of the crater, its full length is only visible on radar images after applying $m-\chi$ decomposition showing the northern end of the lineament extending halfway onto the floor of an adjacent crater lying north of Cabeus B.

Another distinct feature, present on the inner southeastern wall of Cabeus B, is a debris avalanche possibly formed from dislocation and successive downward sliding of debris along the slope of the inner crater wall. It lies on the southern visible tip of the inferred lineament and is only faintly visible on optical images despite falling in the illuminated portion of the lunar surface.

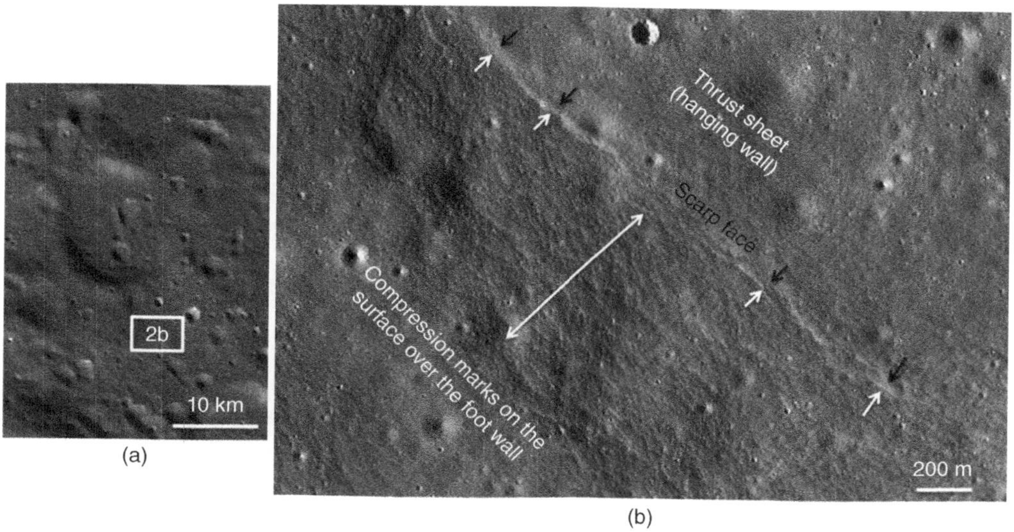

Figure 12.2 Compression marks on scarp segment. (a) Wide angle camera image of the scarp region. (b) Narrow angle camera image of a segment of the inferred scarp marked in (a) showing compression marks on the surface of the footwall of the thrust fault.

The southeastern edge of the feature has a relatively high CPR value in the range of ~1.5–2 suggesting the presence of rough surface within this portion of Cabeus B. The formation of this debris avalanche could possibly be due to the dislocation of debris from movement along the inferred lineament. The possibility of the avalanche being triggered due to impact events nearby is unlikely since the slumping of the debris is present only in a small region of the inner wall of Cabeus B and such a feature is absent elsewhere in other equally sized or relatively larger craters present nearby. Furthermore, other similar sized craters in the South Polar Region show slumping of debris in single and multiple blocks after impact over the entire inner walls of the craters leading to terraced formation. Such type of terracing, however, is absent within the inner walls of Cabeus B.

Also, the possibility of the dislocation and successive sliding down of the material within the avalanche region due to a smaller meteorite impact can also be ruled out since there is no initial impact feature or depression present, which would trigger the subsequent slumping of debris in this region of Cabeus B. It can be seen on the m–χ decomposition image that the orientation of the inferred lineament is almost perpendicular to the debris avalanche. Further southward extension of the lineament could exist and might be passing underneath the visible southern end of the avalanche. The presence of this southward extension of the lineament can be validated using deeper penetrating radar sensors to reveal the extension of the lineament underneath the avalanche region and beyond.

On the southeast region of Cabeus B, Cabeus A crater also shows another segmented lobate scarp lying around 82°26′S and 44°5′W approximately. This feature can also be seen on m–χ decomposition image of the region as a segmented, curvilinear formation. Northwest of the identified scarp lies another short-curved segment of the scarp not visible on optical images since it lies in the shadowed portion. This scarp has formed on the inner wall of a crater and the thrust sheet of the scarp rides up the slope of the inner crater wall. Scarp complexes often have an echelon stepping segments in parallel or sub-parallel orientations. The inferred fault scarp in Figure 12.4c is also suggestive of this type of an echelon fault system.

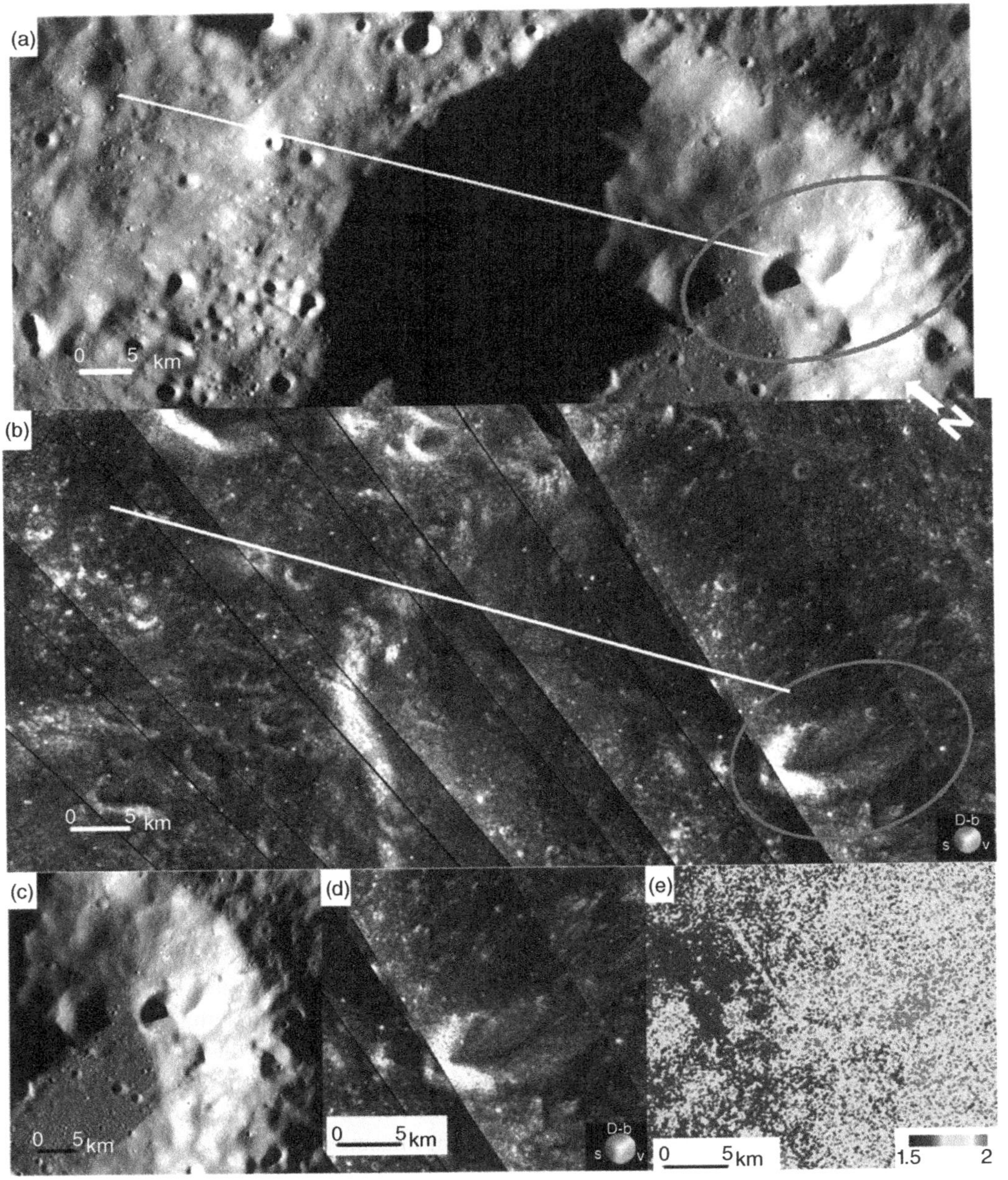

Figure 12.3 Lineament and debris avalanche within Cabeus B (centered at 82°18′S and 54°12′W). (a) Wide angle camera image of Cabeus B crater and surrounding region with the inferred lineament (under the white line). (b) Narrow angle camera image of the scarp region marked in a (white arrows pointing at the scarp foot wall; black arrows mark the hanging wall). (c) Wide angle camera image of the debris avalanche. (d) $m-\chi$ decomposition of the region of the debris avalanche showing volume scattering indicative of surface roughness. (e) CPR image of the region coinciding with the volume scattering indicative of high surface roughness.

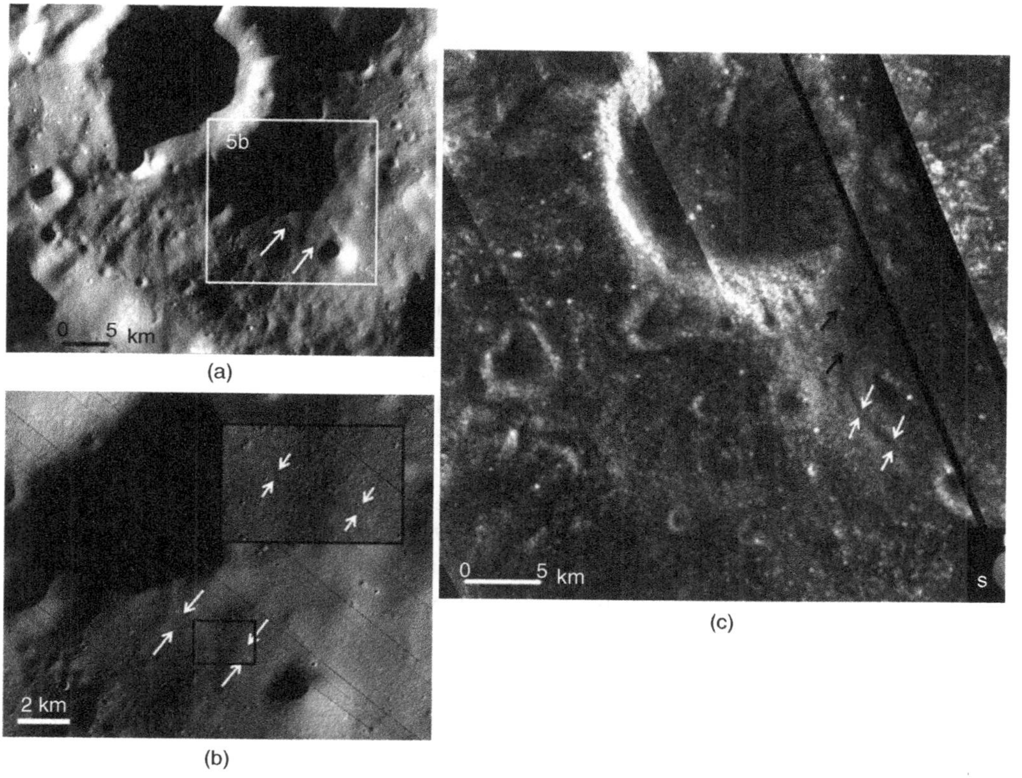

Figure 12.4 Scarp within Cabeus A crater (82°26′S and 44°5′W). (a) Wide angle camera image of the region of the scarp (white arrows mark the visible portion of the scarp within the crater). (b) Narrow angle camera image of the scarp region marked with white boundary in (a). A section of the scarp (inset). (c) $m-\chi$ decomposition of the region in (a), showing another segment of the scarp hidden in (a) (black arrows).

Crustal tectonic activity therefore helps to demarcate regions experiencing physical and geological stress within the planetary interior. The geological activity and the related geodynamics lead to the formation of unique features suitably addressed as the surface manifestations of underlying subsurface perturbations. The formation of such tectonic features derived from perturbations along faults is suggestive of the recent crustal movement of the lunar body. Also, the shallow (<100 km depth) moon-quakes recorded on Apollo data could have resulted due to activity along such type of tectonic deformations present at different locations on the lunar surface [13]. The inferred crustal dislocation features reported in this study suggest that the lunar interior is seismically active and that further geological perturbations as such can be expected.

Furthermore, the use of radar data can reveal hidden tectonic land deformations possibly formed due to disturbances along the subsurface faults. Application of $m-\chi$ decomposition technique on Mini-SAR data effectively helps to demarcate fault lines in shadowed as well as illuminated regions underneath the lunar regolith due to the unique capability of microwaves to penetrate the regolith, completely independent of solar illumination. Therefore, using SAR data and the derived parameters, several complex lineament systems portraying the tectonic peculiarities of a region can be identified. A finer radar sensor will further help to elucidate the formation of tectonic structures hidden in the permanently shadowed polar regions as well as the far-side highlands.

Study of morphological features in Schrödinger basin

- Study of mineralogical characteristics of the basin with reference to landing site selection.
- Identification of landing sites based on the geological observations and mineralogical analysis carried out within the area to understand the basin formation process.

Schrödinger basin is a large peak ring impact basin having a diameter of ~312 km situated at the far side of South Pole centered at 75°S and 130°E. The basin is considered to be the second youngest basin and it lies within the oldest and largest basin South Pole Aitken. The crater floor lies 4–5 km beneath the crater rim and the inner peak ring having a diameter of 125 km rises 1–2.5 km above the basin floors. The basin is estimated to be early Imbrium in the age that makes it the youngest basin with a well-preserved rim, basin floors, and wall terraces. This age marks the last stage of multi-ring basin-forming events and heavy meteoritic impact on the moon.

Based on morphological features and mineralogical findings, four locations were selected as landing sites. The first landing site (LS1) centered at 75.213°S and 138.330°E lies close to the dark halo crater, which is considered to have depositions of dark mantle materials. The second landing site (LS2) is the region having terraces in the crater rim centered at 79.001°S and 146.018°E. Landing site 3 (LS3) lies around the peak ring of the crater basin at 75.62°S and 13.084°E. The fourth landing site (LS4) is the smooth plain region of the basin centered at 74.139°S and 135.183°E, which consists of impact melt or early-stage volcanic materials erupted as a result of floor fractures and emplaced after basin formation. The landing site is chosen based on desirable conditions including surface slopes ≤15°, presence of boulder with size less than 0.5 m, no shadow cast for long durations, and sunlit for at least 14 days in the region (Figure 12.5).

A number of datasets were combined using the ArcGIS software to study the morphology of the basin. Lunar Reconnaissance Orbiter Camera–wide angle camera having a uniform resolution of 100 m pixel^{-1} and Lunar Reconnaissance Orbiter Camera–narrow angle camera with 0.5 m pixel^{-1} resolution have been used for the detailed study of a suitable landing site for future exploration on

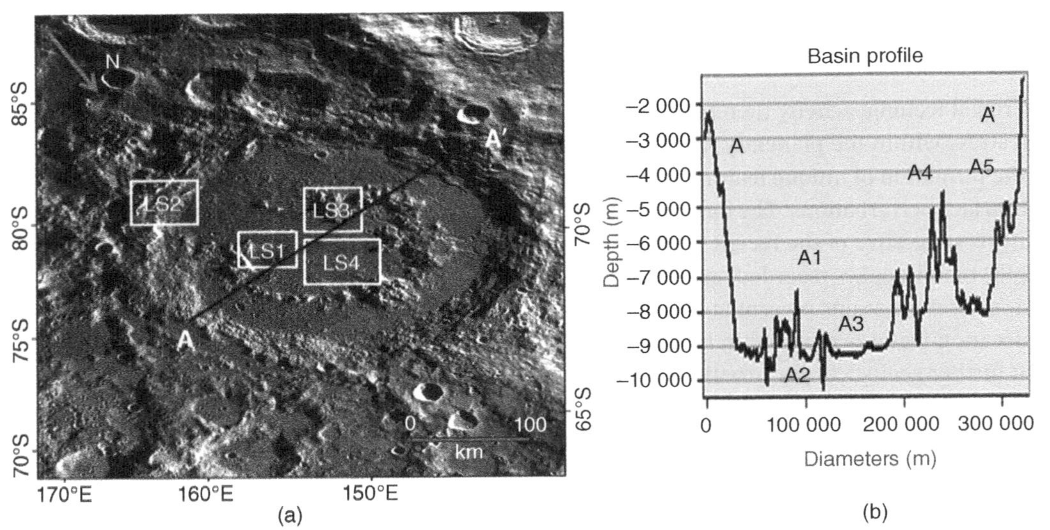

Figure 12.5 (a) LROC-Wide angle camera image of the Schrödinger basin depicting the suggested landing sites LS1, LS2, LS3, and LS4, respectively. (b) The basin profile graph using LOLA DEM that highlights A1 and A4 as crater peak rings and A2 as dark halo crater. A3 are the pyroclast deposits ideal for future landing and A5 is the terrace structure formation on the rim of the basin.

the moon. Lunar Orbiter Laser Altimeter (LOLA) South Pole Digital Elevation Model (DEM) with spatial resolution 30 m pixel^{-1} was used to study the topographic profile of the crater basin. For the study of possible landing sites and flatland nature of the Schrödinger basin, DEM derived from Chandrayaan-1 Terrain Mapping Camera (TMC) DEM with a ground resolution of 5 m and swath coverage of 20 km from an orbit of 100 km was used. The mineral profiling was carried out using Chandrayaan-1 Moon Mineralogy Mapper (m3) level 2 hyperspectral data. M3 has the spectral range from 430 to 3,000 nm over 85 spectral channels. The level 2 data product is the reflectance data derived from level 1B radiance data, which was calibrated to reduce the noise and artefacts in the dataset.

12.1 Landing Site Selection Based on Morphology and Mineralogy of the Basin

The LS1 region is the area around the dark halo crater where dark mantle materials are deposited. The fracture cuts through the crater in North–South direction as seen in Figure 12.6a.

The second landing site LS2 is proposed to be the melt pond around the basin rim with well-developed terrace walls. The region is situated in the northwest direction of the basin that is a promising target for South Pole Aitken basin material. At the time of impact event, the melt is either deposited in the form of central melt sheets or trapped as melt ponds on the inner walls of the crater basin as shown in the white box in Figure 12.7.

The third suggested landing site LS3 lies in the lower Schrödinger smooth plain region close to the peak ring structure. The LS3 region is interpreted to consist of impact melt deposits as a result of volcanic eruption and its further deposition after basin formation. The mineralogical study by Shankar proposed the region to have a mixture of mafic materials and anorthositic rocks. This may

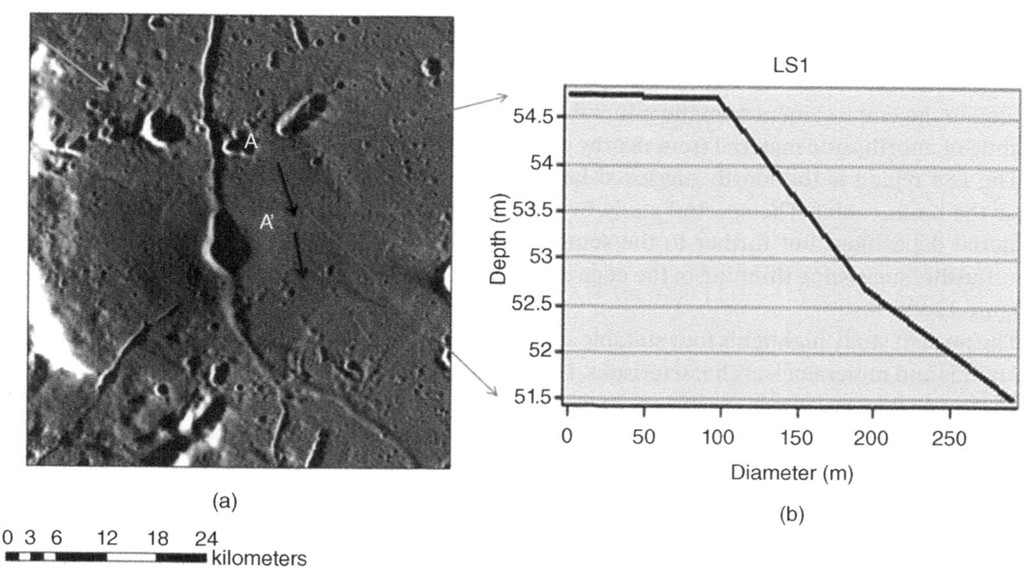

Figure 12.6 (a) LROC-wide angle camera showing landing site 1 wherein segment AA′ marks the flat basin. (b) Basin profile graph depicting the flat basin using Chandrayaan-1 TMC DEM (TMC_NRN_20090529T055154_1_AN) data of a segment bb′ from (a).

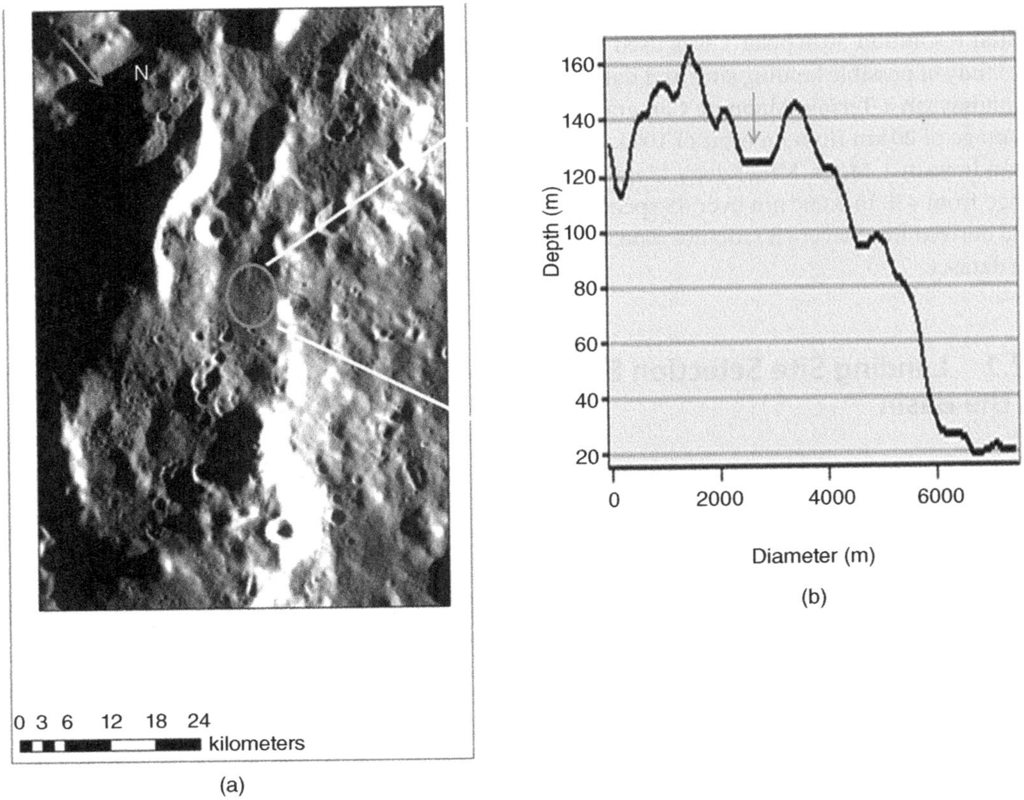

Figure 12.7 (a) LROC-wide angle camera image of the landing site marked with a circle. (b) Basin profile graph of the marked region using Chandrayaan-1 TMC DEM data (TMC_NRD_20090528T124855_1_AN) depicting the flat region in the basin rim.

be due to the emplacement of mafic minerals at the time of volcanic eruptions with the overlying mantle of anorthositic material from nearby crater impacts (Figure 12.8).

The LS4 region is the fourth suggested landing site, which lies in dark plain material region near the basin center. The spectral study by Shankar highlights that the region has more mafic material deposition. But further to the south of LS4, there is a mixing of mafic minerals with anorthosite, suggesting thinning of the edge of the dark mantle region as a result of crater impacts (Figure 12.9).

The present study highlights four suitable sites for future exploration that have different geological trends and mineralogical characteristics. Landing Site 1 (LS1) was seen having the deposition of dark mantle material of pyroclastic origin that will help in the study of final stage of lunar volcanism. Landing site 2 (LS2) was selected within the crater rim with well-developed terrace and melt pond formations. The region shows older deposits than the Schrödinger basin, making it a suitable site for the study of basin formation. The lowermost terrace, in particular, might have preserved around 20–30% quenched SPA impact melt and, thus, represents a potential destination for further robotic and human investigations. Landing site 3 (LS3) marked the presence of early-stage volcanic deposits, and collecting samples from the region may provide us with information about the peak ring formation. Landing site 4 (LS4) lying in the dark mantle region had numbers of ghost craters highlighting the deposition of draped lava around the craters.

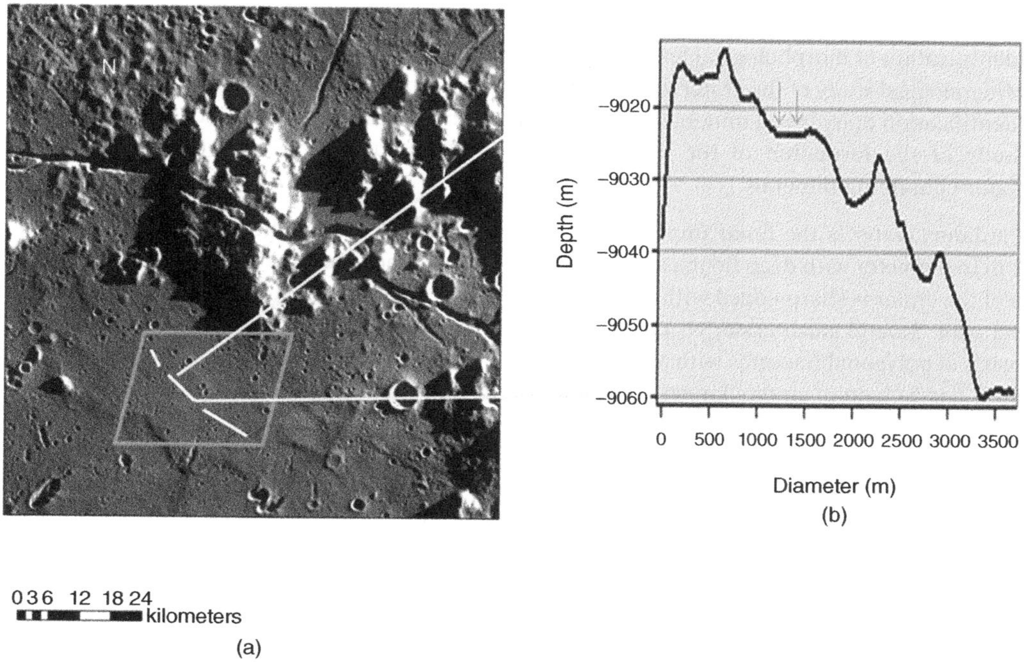

Figure 12.8 (a) LROC-wide angle camera image showing landing site marked in white arrows where region is smooth plain. (b) Basin profile graph of demonstrating the flat land in black arrows using LOLA DEM.

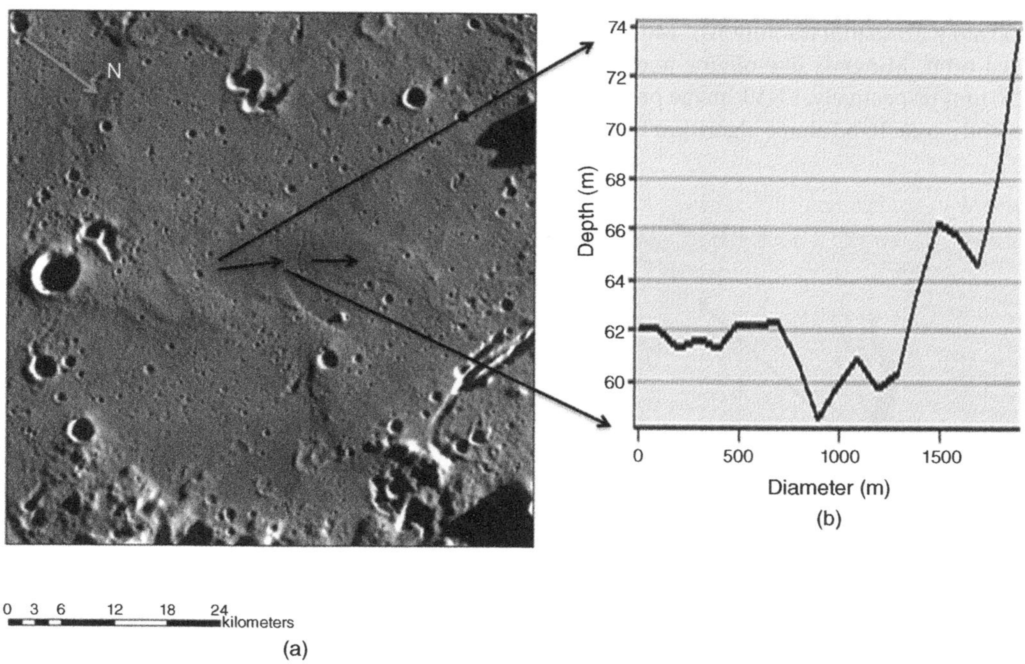

Figure 12.9 (a) LROC-wide angle camera image of the landing site marked in black lines showing dark plain material deposition. (b) basin profile graph of a part of the image indicates the flatland using Chandrayan-1 TMC DEM data (TMC_NRN_20090529T055154_1_AN).

It can be inferred in Schrödinger basin

- Identification of morphological features within the Cardanus crater.
- Mineralogical study of the crater with reference to the formation of fractures.
- Identification of hydroxyl minerals deposited within the fractures of crater floor.
- Study of the formation of the crater fractures and its correlation with the deposition of water-containing minerals.

Cardanus crater is the lunar impact crater centered around 13.2°N and 72.4°W. The crater is 50 km in diameter with deep floors, a complex central peak, and extensive wall terraces. The outer rim of the crater is sharp-edged with the hummocky outer rampart and terraces along parts of the inner wall. Several small craterlets can also be seen on the surface of the crater floor. The crater consists of polygonal fractures with a deep floor, central peak, and terraces along parts of the inner wall. It has a chain of craters known as Catena Krafft that connects it to the Krafft crater located in its north. The crater was found to be of late Imbrium period having an age of about 3.5 billion years (Figures 12.10–12.12).

For the morphological study of the crater, LROC-wide angle camera and LOLA DEM images have been used. Wide angle camera (WAC) is a pushbroom imager that works on both multispectral and monochrome modes. Global WAC mosaic image was used for crater age determination using crater stat tool 2.0. Craters greater than 250 m diameter that surround the Cardanus crater were mapped using Arc GIS 10.1 to prevent the mapping of secondary craters. The craters were binned with pseudo-log function in crater stat tool. Chronology function and production function was used to determine the crater age through crater size frequency distribution plot. LOLA data from lunar Reconnaissance orbiter has been used to study the elevation of crater.

For mineralogical analysis, photometrically and thermally corrected Level-2 data products of Chandrayaan-1 Moon mineralogical Mapper have been used. It has a spectral range from 460 to 3,000 nm with spatial resolution of 140 m pixel^{-1} from 100 km orbit and 280 m pixel^{-1} from 200 km raised orbit. Minerals like olivine and plagioclase have strong absorption band near 1,000 and 1,250 nm, respectively. ENVI image processing software is used for the mineralogical mapping of the study area. Band shape algorithm has been done to delineate the abundance of minerals in the

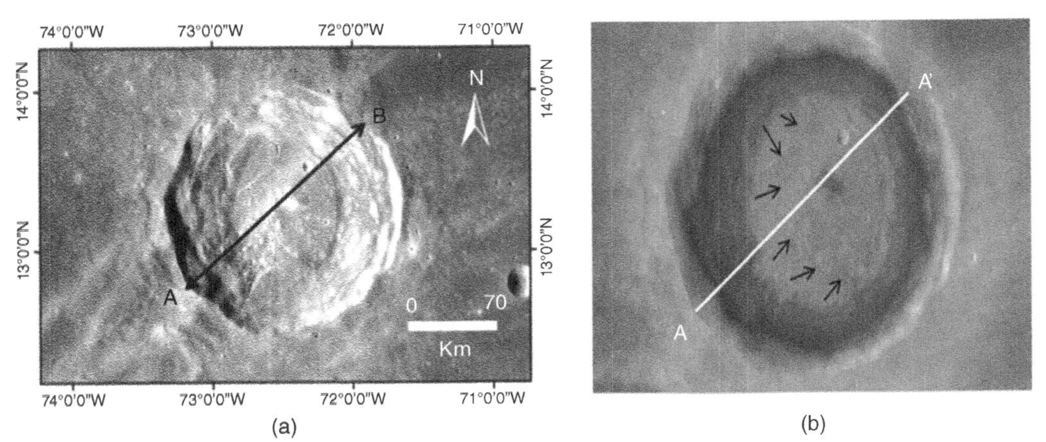

Figure 12.10 (a) LROC-wide angle camera image of the study area. (b) LOLA image of Cardanus crater showing fractures.

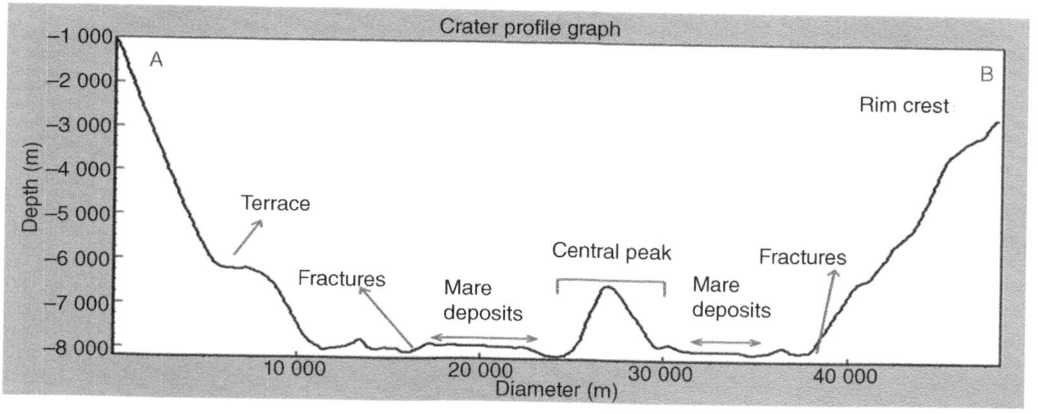

Figure 12.11 Crater profile graph drawn from LOLA merged Digital Elevation model fractures in the periphery of the crater and mare deposits around the central part of the crater. The graph also features the terracing within the inner crater wall.

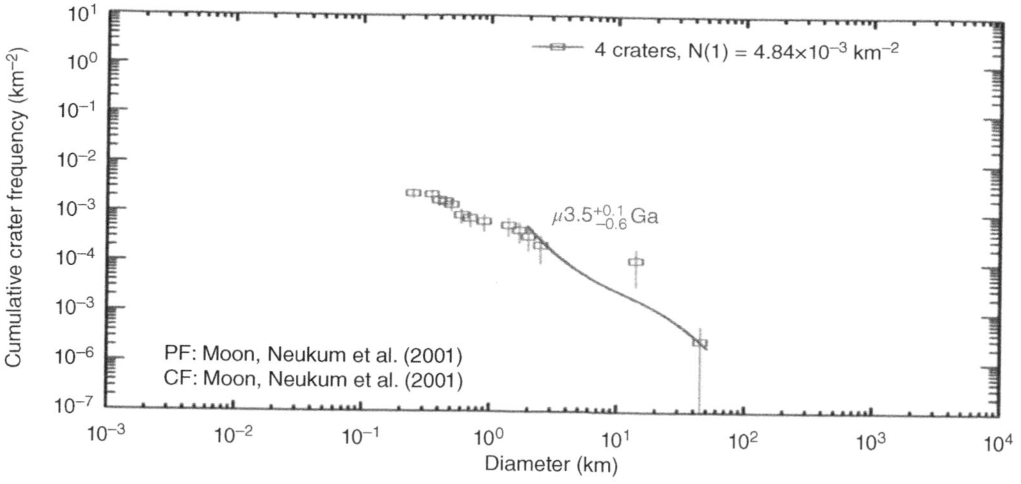

Figure 12.12 Crater age dating based on crater count using Crater stats 2.0 tool.

study area. The algorithm includes band strength (bs), band curvature (bc), band tilt (bt), and band ratio (br). These are calculated at a specific wavelength to distinguish specific minerals in false color composite map.

$$\text{bs} = 1009/750\ \text{nm}$$

$$\text{bc} = (750/910 + 1009/910)\ \text{nm}$$

$$\text{bt} = 910/1009\ \text{nm}$$

$$\text{br} = 2018/1009\ \text{nm}$$

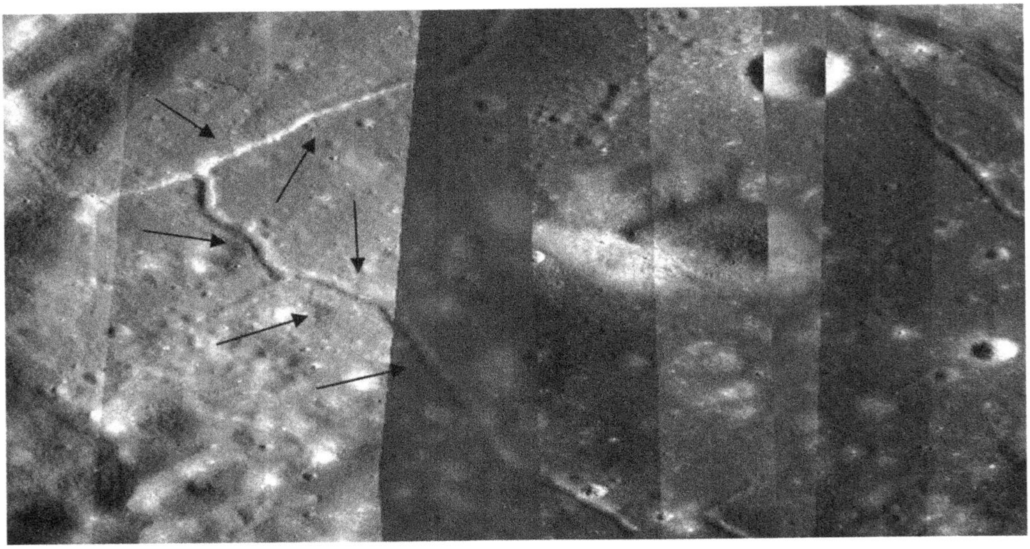

Figure 12.13 LROC-wide angle camera image showing fracture formation within the crater.

12.2 Morphological Study of the Crater

12.2.1 Fractures on Crater Floor

The fractures present within the crater floor are the surface manifestation of a class of shallow crustal intrusions in which magma-filled dikes rise vertically toward the surface from the mantle region. It laterally forms sill or laccolith that creates overburden pressure on the periphery of the crater rim. The crater wall and rim deposits restrict the lateral growth of the sill, which forces it to grow vertically and form piston-like structures, uplifting the periphery of the crater floor (Figure 12.13).

12.2.2 Central Peak

The central peak is an elevated feature formed by an impact on the moon surface. Cardanus crater has a group of central peaks near the mid-point. The central peak has olivine bearing units identified based on absorption band characteristics (Figure 12.14).

12.2.3 Terraced Walls

Complex crater walls are commonly broken up into a series of terraces that are relatively flat ledges, offset by scarps facing the center of the crater, which steps down from the crater rim to the floor. The width of the terraced wall increases from the crater floor outward, with the largest and most defined terrace under the crater rim (Figure 12.15).

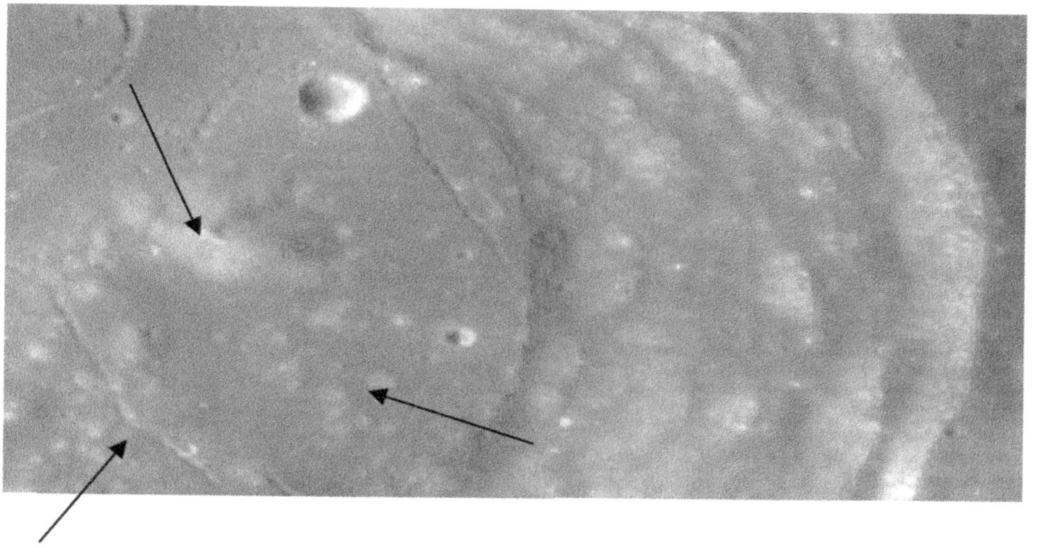

Figure 12.14 Presence of central peak within the crater.

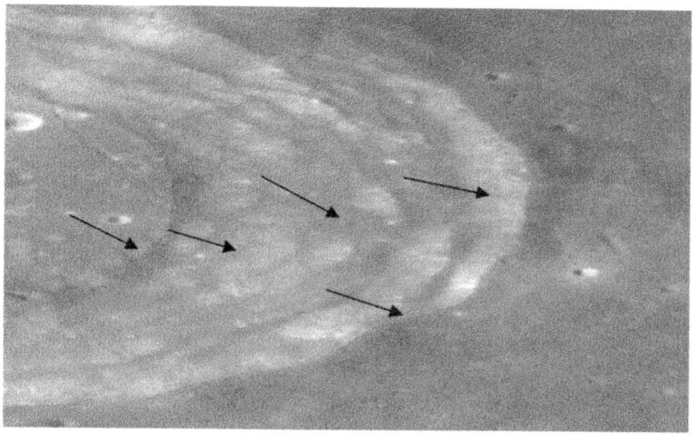

Figure 12.15 LROC-narrow angle camera image showing terrace present at the rim of crater.

12.2.4 Rampart Crater

Rampart craters are a specific type of impact crater, which are accompanied by distinctive fluidized ejecta features. They show a lobate outer margin, as if material moved along the surface, rather than flying up and down in a ballistic trajectory. Cardanus crater shows partial rampart features (Figure 12.16).

12.2.5 Formation of Catena Krafft

Catena Krafft is the chain of craters that connects Cardanus crater to Krafft crater in the north (Figure 12.17).

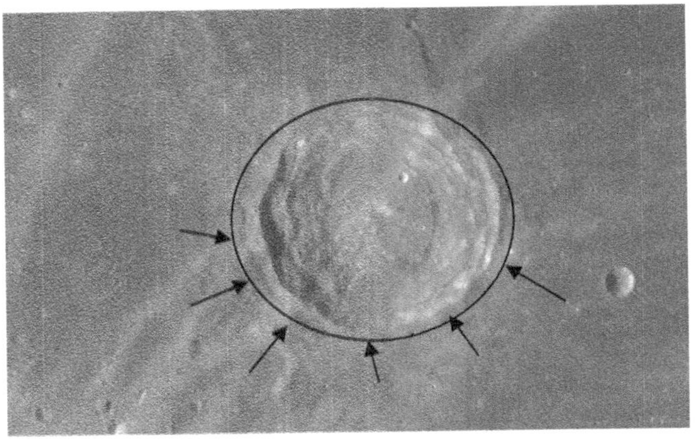

Figure 12.16 LROC-wide angle camera image showing partial rampart structure around the crater.

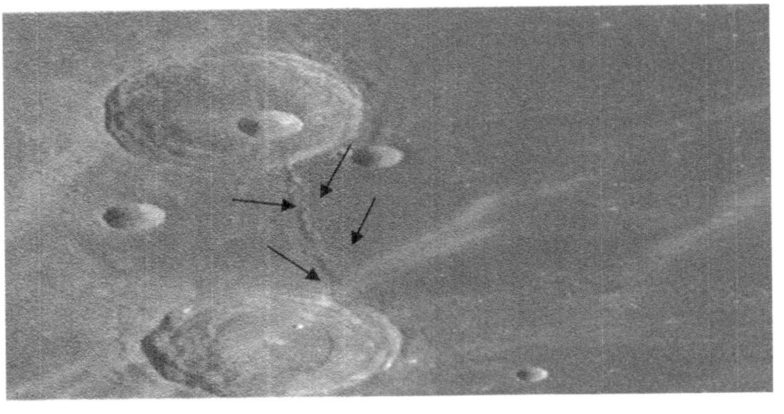

Figure 12.17 Catena Krafft between Cardanus crater and Krafft crater.

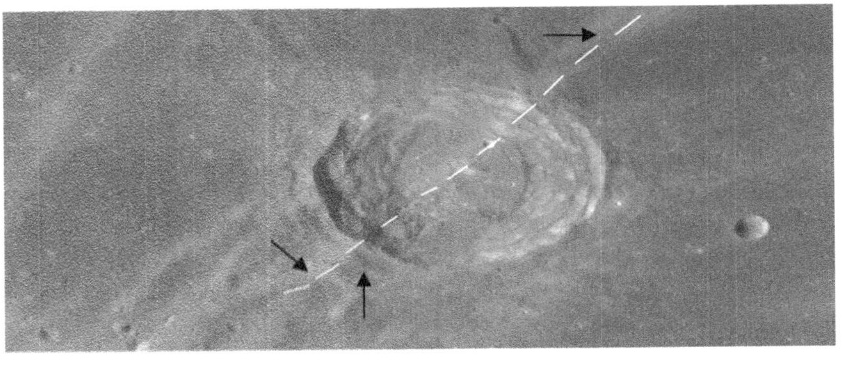

Figure 12.18 LROC-wide angle camera image showing rays around the crater.

12.2.6 Rays Around the Crater

Cardanus crater is found to have the depositions of the impact of Glusko crater that can be seen in the form of rays around the crater (Figure 12.18).

12.3 Morphological Variations in Light of Mineralogy of the Study Area

The fractured floor crater has olivine rich area, which was delineated through the ratio of 2018/1008 nm. RGB composite shows high-band tilt values that appear in green color indicating the presence of olivine-rich lithology. It has a strong absorption band near 1000 nm and there is no significant absorption near 2000 nm. Around the fractured floors, band curvature values are high, which probably indicates the presence of low-calcium pyroxene that has high Fe-Mg components. Fe–Mg components show strong absorption near 2000 nm and weak absorption near 1000 nm. In some areas of the fractured floor, FCC shows some patches of olivine-bearing lithology associated with the hydration features. These olivine and OH features have been identified based on its absorption feature at 1070 and 2816 nm, respectively. The fractured floor crater that is to the south of the crater has high calcium pyroxenes with their band 1 absorption features near 940 and 985 nm and band 2 absorption features near 2045 and 2135 nm, respectively.

The fractures identified on the floor of the crater were found to be rich in silicate minerals consisting of olivine associated with hydroxyl ions. The presence of hydroxyl ions was identified from the absorption peak around 2800–3000 nm in the reflectance spectra. The region was also found to be rich in olivine having absorption near 1000 nm that was delineated by the ratio of the wavelength at 2018/1008 nm on gray scale (Figure 12.19). The abundance of olivine in the fracture describes it to be an excavation of the lunar mantle at the time of impact that formed the crater floor trapped in the oxygen-rich silicate minerals. Craters containing fractures are more prone to the formation of such hydroxyl-containing minerals through sputtering process. Fractures on the crater floor result in dislocation of the crystal lattice of the mineral. The bond length and bond strength become irregular in the distorted crystal lattice. These fractures trap protons and get bonded with the abundant oxygen-forming hydroxyl minerals. It is considered that protons mostly interact with the basaltic silicate minerals and other oxygen-bearing minerals in the lunar basalt (Figure 12.20).

The presence of silicic minerals in the fractures within the crater can be a suitable site for proton capture and hydroxyl ion formation. Our study highlights that the fractures present in the Cardanus crater are a rich source for hydroxyl-containing olivine mineral. Due to the disturbance in the mineral lattice, protons from solar radiation hitting the fractures get trapped and bonded with olivine-forming hydroxyl mineral. Further, dike emplacement and sill formation in the mantle region resulting in fracture formation might lead to peripheral upliftment around the crater rim with subsidence in the central region of the crater. This can transform the crater into a reservoir for hydroxyl-/water-containing minerals.

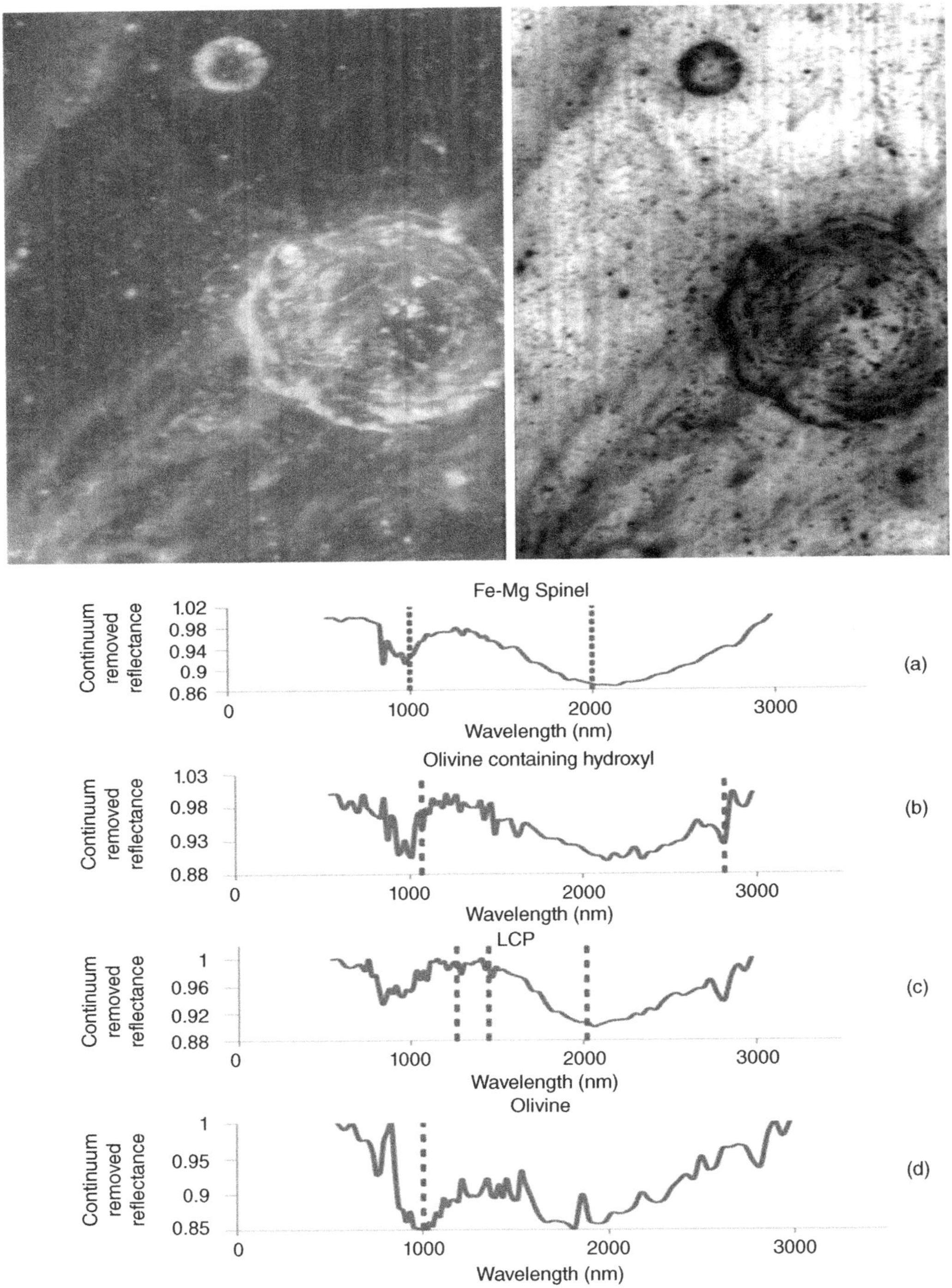

Figure 12.19 RGB Composite of bs, bc, and bt in the study area. Gray scale image of the study area based on 2018 nm/1008 nm. (a–d) showing the continuum removed reflectance spectra of various minerals within the floor fractured crater. (a) Shows strong absorption near 2000 nm. (b) Shows absorption bands between 1000 and 2000 nm. (c) Shows olivine mineral with strong absorption feature at 1000 nm. (d) Mark the olivine containing hydroxyl mineral with strong absorption feature at 1070 and 2816 nm.

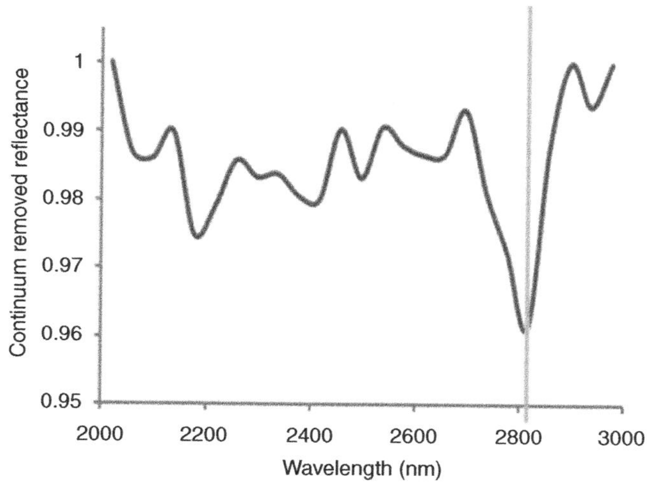

Figure 12.20 Continuum removed reflectance spectra of hydroxyl olivine showing reflectance at 2800 nm in the fractured region of crater basin.

References

1 Anand, M., Taylor, L.A., Nazarov, M.A. et al. (2004). Space weathering on airless planetary bodies: Clues from the lunar mineral hapkeite. *Proceedings of the National Academy of Science of the USA* 101: 6847–6851.

2 Binder, A.B. and Gunga, H.-C. (1985). Young thrust-fault scarps in the highlands: Evidence for an initially totally molten moon. *Icarus* 63: 421–441.

3 Burke, D.J., Dukes, C.A., Kim, J.H. et al. (2011). Solar wind contribution to surficial lunar water: Laboratory investigations. *Icarus* 211 (2): 1082–1088.

4 Carter, L.M., Campbell, D.B., and Campbell, B.A. (2011). Geologic studies of planetary surfaces using radar polarimetric imaging. *Proceedings of the IEEE* 99: 770–782.

5 Condie, K.C. (1997). *Plate Tectonics and Crustal Evolution*, 4e, 134–136. Elsevier Ltd.

6 Crider, D.H. and Vondrak, R.R. (2000). The solar wind as a possible source of lunar polar hydrogen deposits. *Journal of Geophysical Research: Planets* 105 (E11): 26773–26782.

7 Fa, W., Wieczorek, M.A., and Heggy, E. (2011). Modeling polarimetric radar scattering from the lunar surface: Study on the effect of physical properties of the regolith layer. *Journal of Geophysical Research* 116: 1–33.

8 Lucey, P., Korotev, R.L., Gillis, J.J. et al. (2006). Understanding the lunar surface and space-moon interactions. *Reviews in Mineralogy & Geochemistry* 60: 83–219.

9 McCord, T.B., Taylor, L.A., Combe, J.P. et al. (2011). Sources and physical processes responsible for OH/H2O in the lunar soil as revealed by the Moon Mineralogy Mapper (M3). *Journal of Geophysical Research: Planets* 116 (E6): 1–22.

10 Melosh, H.J. (2011). *Planetary Surface Processes* (ed. F. Bagenal, D. Jewitt, C. Murray, et al.), 14–16. New York: Cambridge University Press.

11 Mukherjee, S. and Singh, P. (2014). Application of m-χ decomposition technique on mini-SAR data to understand crater and ejecta morphology. *IEEE Geoscience & Remote Sensor Letters.* 12: 73–76.

12 Patterson, G.W., Raney, R.K., Cahill, J.T.S., and Bussey, D.B.J. (2012). Characterization of lunar crater ejecta deposits using m-chi decompositions of mini-RF data. *Proceedings of European Planetary Science Congress* 7: 731.
13 Watters, T.R., Robinson, M.S., Beyer, R.A. et al. (2010). Evidence of recent thrust faulting on the Moon revealed by the Lunar Reconnaissance Orbiter Camera. *Science* 329: 936–940.

Further Reading

Mukherjee, S. and Singh, P. (2014). Investigation of tectonic processes in the lunar South Polar Region using Mini-SAR and other data. *Frontiers in Earth Science* 2: 97–112.

Mukherjee, S. Roy, N. Singh, P., and Singh, D. (2018). Mineralogical analysis of the fractured floor region within Cardanus crater. *Lunar and Planetary Science Conference (LPI Contrib. No. 2083),* The Woodlands, TX (19–23 March 2018). https://www.hou.usra.edu/meetings/lpsc2018/.

Nakamura, Y., Latham, G.V., Dorman, H.J. et al. (1979). Shallow moonquakes: Depth, distribution and implications as to the present state of the lunar interior. *Proceedings of Lunar Science Conference* 10: 2299–2309.

Raney, R.K., Cahill, J.T.S., Patterson, G.W., and Bussey, D.B.J. (2012). The *m-chi* decomposition of hybrid dual-polarimetric radar data with application to lunar craters. *IEEE International Geoscience and Remote Sensors Symposium.* 117: 5093–5096.

Saran, S., Das, A., Mohan, S., and Chakraborty, M. (2012). Study of scattering characteristics of lunar equatorial region using Chandrayaan-1 Mini-SAR polarimetric data. *Planetary and Space Science* 71: 18–30.

Schultz, R.A., Hauber, E., Katternhorn, S.A. et al. (2010). Interpretation and analysis of planetary structures. *Journal of Structural Geology* 32: 855–875.

Solomon, S.C. and Chaiken, J. (1976). Thermal expansion and thermal stress in the Moon and terrestrial planets: Clues to early thermal history. *Proceedings of the Lunar Science Conference* 7: 3229–3243.

Solomon, S.C. and Head, J.W. (1979). Vertical movement in mare basins: Relation to mare emplacement, basin tectonics and lunar thermal history. *Journal of Geophysical Research* 84: 1667–1682.

Solomon, S.C. (1977). The relationship between crustal tectonics and internal evolution in the moon and Mercury. *Physics of the Earth and Planetary Interiors* 15: 135–145.

Spudis et al. (2010). Initial results for the north pole of the moon from Mini-SAR, Chandrayaan-1 mission. *Geophysical Research Letters* 37: 1–6.

Varatharajan, I., Srivastava, N., and Murty, S.V. (2014). Mineralogy of young lunar mare basalts: Assessment of temporal and spatial heterogeneity using M3 data from Chandrayaan-1. *Icarus* 236: 56–71.

Watters, T.R. and Johnson, C.L. (2010). *Planetary Tectonics* (ed. T.R. Watters and R.A. Schultz), 121–182. Cambridge University Press.

Watters, T.R., Robinson, M.S., Banks, M.E. et al. (2014). Global distribution of lobate scarps on the moon: Implications for the current stress state. *45th Lunar and Planetary Science Conf. Abstract #* 2163.

Watters, T.R., Robinson, M.S., Banks, M.E. et al. (2012). Recent extensional tectonics on the Moon revealed by the Lunar Reconnaissance Orbiter Camera. *Nature Geoscience Letters* 5: 181–185.

Watters, T.R., Robinson, M.S., Beyer, R.A. et al. (2009). A new look at formation and timing of thrust fault scarps on the Moon. *American Geophysical Union, Abstract #U31A-0013* Fall.

Wells, K.S., Campbell, D.B., Campbell, B.A., and Carter, L.M. (2010). Detection of small lunar secondary craters in circular polarization ratio radar images. *Journalof Geophysical Research*. 115.

Williams, N.R., Watters, T.R., Pritchard, M.E. et al. (2013). Fault dislocation modeled structure of lobate scarps from Lunar Reconnaissance Orbiter Camera digital terrain models. *Journal of Geo. Research: Planets* 118: 224–233.

Wilson, L. and Head, J.W. (2018). Lunar floor-fractured craters: Modes of dike and sill emplacement and implications of gas production and intrusion cooling on surface morphology and structure. *Icarus*.

13

Remote Sensing of Mars

Mars Orbiter Probe (Mangalyaan) was launched on 5 November 2013 by the Indian Space Research Organisation. It is India's first interplanetary mission (Figure 13.1).

The study of Mars was carried out in its different morphological units. There was presence of surface water in the geological past, which is recorded as fluvial geomorphological features on Mars. Location of these features is region specific and some equatorial regions are showing presence of desert-like sandy dunes on Mars. Remnants of old saline lakes were found to be endorheic playa in the equatorial region. The interpretation of Martian data gives initial information of a life support system. Besides water and life, the focus is on the location for safe landing of Martian spacecraft in a future mission.

After volcanic eruption, the Martian surface develops some depressions, which change into lakes, and we needed to study this in more detail. Some of these depressions and its surrounding morphological features were actually filled with water, and it is well established now that Mars was hydrologically more active than previously thought. Remnants of old fluvial geomorphological features including valley shapes were discovered within the equatorial region of Mars. One of the impact craters, named Huygens crater, further proves this concept by the presence of a topographically flat basin toward the north. In general, all the impact craters and volcanic emission basin can be divided into two geological units on the basis of their morphological differences. A number of morphological units inferred on Mars were wrinkled ridges, dikes, different types of polygonal desiccation patterns, and craters with layered ejecta blankets. They were observed within these units. Morphological variability including lobate shapes, ejecta mobility, and depth of the diameter ratio of craters within these units which suggest climate-controlled fluctuations of volatiles and multiple episodes of resurfacing. Geologically, it is possible to reconstitute and infer a possible stratigraphy of Mars after integrating basin morphology with cumulative size frequency distribution (CSFD) analysis. Detailed investigation of the terrain suggests that the basin is an ephemeral, endorheic playa that underwent several geodynamic processes and is a geologically rich terrain for future exploration missions.

Some fluvial geomorphological features were inferred, which includes outflow channels, valley networks, and fluvio-lacustrine basins observed on the Martian surface. It proves the role of water in erosion, sediment transportation, and deposition. A cumulative morphological feature includes valley networks in Southern highlands and Northern lowlands, inter-crater plains, volcanoes, and walls of craters. Its spectral response is identical, but the patterns are distributed only within the equatorial region. The formation of valley networks has been attributed to precipitation, surface runoff, groundwater outflow impact, and subsequent melting of volatiles from magmatic vents.

Presence of fluvio-lacustrine basins has been reported to be hydrologically active on Mars for thousands of years. Valley networks were found in open or closed basins, which may have deposited

Extraterrestrial Remote Sensing and Climate Change, First Edition. Saumitra Mukherjee.

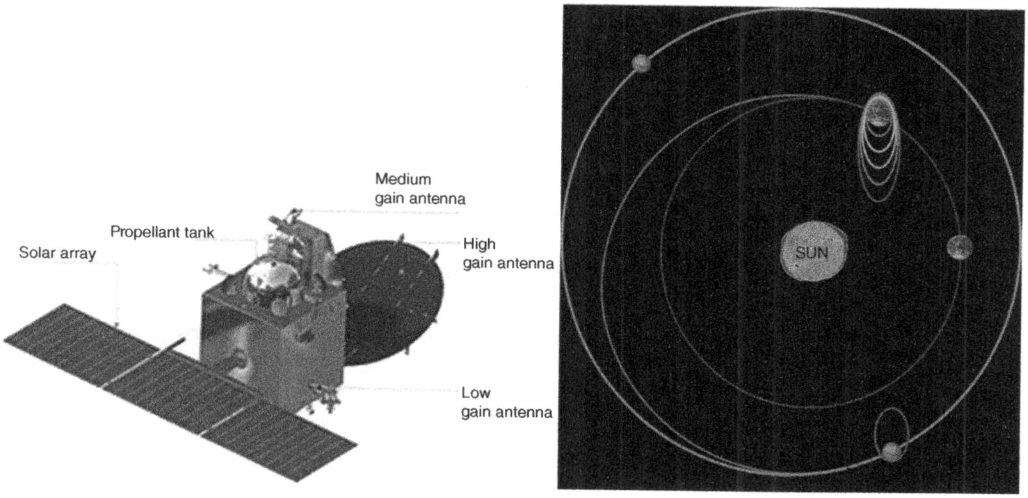

Figure 13.1 Mars Orbiter Mission (MOM) probe schematic diagram. Mars orbiter and its orbit position in space. Source: Indian Space Research Organisation/Wikimedia commons/Public domain.

verve-like materials in closed basins, and less saline basins may have groundwater links. These basins offer insight into past hydrogeological processes on the Martian surface through invading valley networks, morphological features including alluvial fans, deltas, desiccation polygons, etc., and deposited sediments that may be either transported or produced *in situ*. In addition, pounded water provides a conducive environment for harboring life and preserving their biosignatures. Thus, extensive analysis has been carried out on the paleo lakes of Mars and detailed catalogues are available today with their morphological as well as mineralogical profiles.

However, inter-crater depressions, which may also be candidate sites for such lakes to form, have not been included in these catalogues. Topographic lows were found to be present in regionally flat areas of the Martian surface, which were identified as closed basins and may record varying climatic conditions. For instance, varying morphologies and drainage densities of valley networks indicate different hydrological frameworks. This is especially true if these depressions have multiple sources feeding them, which may be part of different valley networks. These regions are in the vicinity of large impact craters, which may have been affected by impact-induced tremors and geothermal Martian water reservoirs or geothermal activity. This may consequently give us diverse geology to explore. The open and closed basin lakes merely contribute about a few meters, respectively, to the total estimate of 5–6 km global equivalent layer on Mars. Some parts of the regional depressions were identified to be filled with water, which justifies that Mars was hydrologically more active than previously thought. Current investigation draws attention toward topographical depressions and its importance to infer paleo environment since they form a substantial component of paleo lake records on Mars. It is pertinent to mention that the changes on Mars are comparable with the fluvial tectonic activities on Earth.

Detailed study has been conducted on the valley networks in the equatorial region of Mars, focusing on areas where valley networks enter topographic depressions. A paleo-drainage network around Huygens crater was examined, and a low topography basin toward its north was found. Detailed morphological analysis of the layered ejecta craters within the basin was carried out to understand the role of volatiles in their formation. It appears that the presence of water molecules in the volcanic emission were the source of the water in vesicular basaltic mass. Morphometric analysis of the craters was done to understand resurfacing events and subsequent modifications occurring over time within the region. Additional morphological investigation such as identification of desiccation polygons was carried out to infer the presence of water in the past. These observations were used to ascertain that the depression was once hydrologically active basins, which estimate its longevity and life cycle.

The global survey of the valley networks was carried out using Mars Color Camera onboard the Mars Orbiter Mission (MOM). It is a medium resolution camera that operates in the visible part of the spectrum. It is able to provide high-resolution localized scenes as well as synoptic views of the planet depending upon the orbit. The images were downloaded from the website https://mrbrowse.issdc.gov.in/MOMLTA/.

Global datasets were for morphological study of the basin including base maps of the Thermal Emission Imaging System–Infrared (THEMIS-IR) Day Mosaic (100 m pix^{-1}), High-Resolution Stereo Camera and Mars Orbiter Laser Altimeter Blended Digital Elevation Model (HRSC MOLA Blended DEM) (200 m pix^{-1}), and Viking Mars Digital Image Model (MDIM2.1) Colorized Global Mosaic (232 m pix^{-1}). The base maps were downloaded from https://astrogeology.usgs.gov/. High-resolution context camera (CTX) (6 m pix^{-1}) and High Resolution Imaging Science Experiment (HiRISE) (0.25–1.3 m pix^{-1}) images onboard the Mars Reconnaissance Orbiter (MRO) for cumulative size frequency distribution and other morphometric analysis were incorporated. Data for the MRO instruments was downloaded from http://ode.rsl.wustl.edu/mars/.

Mars Color Camera PDS level 3 and areo-referenced images were used for observing valley networks. The images were geometrically corrected for slight shifts in their positioning using MDIM2.1 global mosaic. The delineation of drainage network around Huygens crater was carried out using MOLA and HRSC Blended DEM. Initially, the sinks in the Digital Elevation Model were removed using the "Fill" tool. Based on the topographic variations, the "flow direction" and "flow accumulation" were interpreted prior to automatic extraction of the valley networks with the DNR hydromod program on ArcGIS 10.1 software. The basin boundary was demarcated with the help of extracted drainage network and topographic profiles of the basin.

Context camera images were mosaicked for detailed morphological analysis within the region. To determine the model ages of the basin and the layered ejecta craters (craters "P," "R," "S" and "T"; [Figure 13.2] in the Martian region, crater counting was carried out using Context mosaic using Crater [CTX] Tools extension on Arc GIS 10.1. The areas where crater counting has been done are shown in Figure 13.3). First, different subunits were delineated based on regional morphology and relative brightness in THEMIS IR image. Impact craters can be identified based on unique shape of rims, which were not visible but were still highlighted by a pattern of circular wrinkle ridges. To minimize secondary crater contamination, craters with a diameter of less than 100 m and a spatial distribution suggestive of clustering were avoided. For that reason age dating of one of the rampart craters in our study of Crater "Q," (Figure 13.2) was not carried out since the craters over the ejecta blanket of this crater measured below the 100 m diameter benchmark. The age dating was performed using Craterstats2 software (available at http://www.geo.fuberlin.de/en/geol/fachrichtungen/planet/software/index.html). The crater size-frequency distribution (CSFD) plot was made using production function from Ivanov and isochron model coefficients in the Craterstats2 software. It is established that for sample area less than 100 km^2, resurfacing as well spatial variability in the crater pattern can introduce significant differences in the model ages.

CTX images were used to determine diameter, perimeter, and total surface area of the ejecta blanket of the craters labeled "P," "Q," "R," "S," and "T" (Figure 13.2) and were used to calculate other geometric parameters such as oblateness, ejecta mobility, and excavation depth. Typical lobe-shaped morphology and vent ejecta mobility were determined to understand the fluidity of

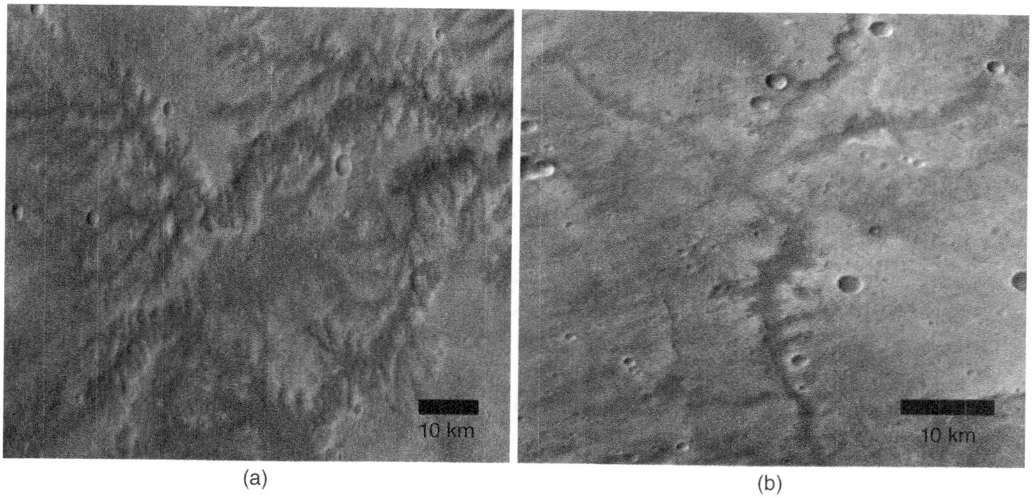

Figure 13.2 Mars Color Camera onboard the Mars Orbiter Mission showing drainage networks around Huygens crater. (a) and (b) show the drainage on the eastern and western side of the Huygens basin.

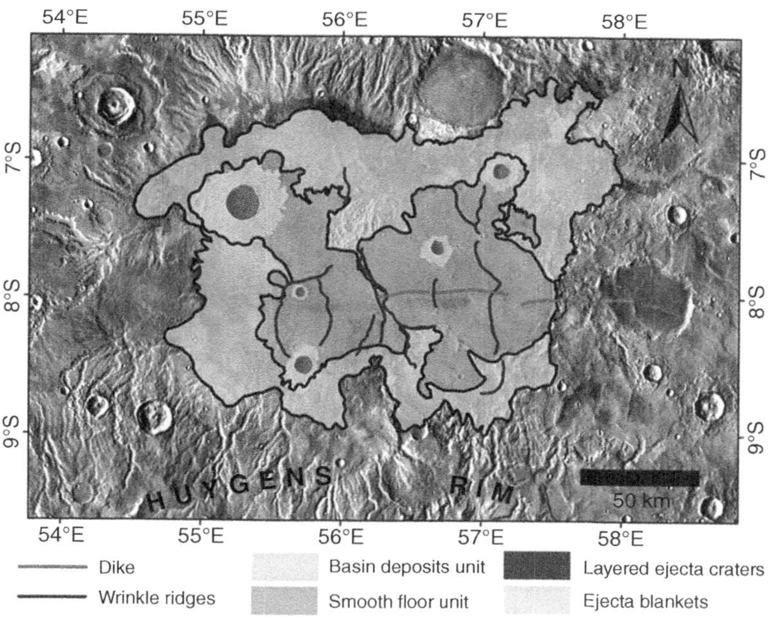

Figure 13.3 Geomorphology map of the basin.

the volatiles present within the basin. In addition to layered ejecta craters, we shortlisted craters from the basin based on their size and location. The study area is outlined in black. The star in the inset shows the location of the study area on Mars. White, red, orange, purple, and brown arrows denote the difference in ejecta morphology of craters P, Q, R, S, and T, respectively.

The study area lies in the Iapygia quadrangle of Mars and extends from around 54.5°E and 6.6°S to 57.1°E and 8.8°S. It is flanked by Schroeter crater in the north and Huygens crater in the south (Figure 13.2). The region was dated to the late Noachian period bounded by middle and early Noachian units on the northwest and southern sides. The study area is surrounded by valley networks on the northern, western, and southern sides. The inlet valleys feeding the basin have been dated back to Noachian-Hesperian boundary. The terrain is basaltic and rich in olivine and high-calcium pyroxene (HCP) minerals. This basin is also close to one of the chloride-bearing terrains. The floor of the basin is marked by the presence of several wrinkle ridges and an extension of the dike network, running through the center of the basin. Volcanic as well as resurfacing from unknown sources has been reported in the open basin lakes on the northern side of the basin.

The global survey of valley networks conducted using Mars Color Camera revealed a dense, dendritic network near the eastern rim of Huygens crater in comparison to the long drainage streams having fewer distributaries near the western boundary (Figure 13.2b). Differences in the morphology of the drainage streams despite their proximity to the apparent source led us to explore these networks in detail.

The basin has been divided into two units: Basin Deposits Unit (BDU) and Smooth Floor Unit (SFU). Basin Deposits Unit constitutes the part of the basin which is not resurfaced. Smooth Floor Unit is marked by the presence of several craters with layered ejecta, wrinkle ridges, and a dike, which runs through the center of the basin.

Based on the morphology, a geomorphological map of the basin (Figure 13.3) was prepared.

SFU forms the center of the basin and covers about 9047 square kilometers. It is shown with relatively darker tone in Figure 13.3. This unit appears smooth in texture with morphology similar to other volcanic terrains reported on Mars. It has numerous small craters of a few hundred meters distributed uniformly throughout, several big craters varying from 3–4 to 51 km in diameter with partially degraded rims, as well as craters with layered ejecta blankets. The ejecta blankets of some of the craters have also been eroded. A dike extending in the east–west direction (Figure 13.3) continues to the outside of the basin to the northeast of Huygens crater. Dust layers and dark mantle overburden partially covering small craters were observed in the central part of the unit. The presence of several wrinkle ridges running in the north-south direction (represented in Figure 13.3) gives the unit an uneven topography.

BDU lies toward the basin periphery and contributes to 10 997 square kilometers of the total basin area. It is shown with relatively lighter shade in the geomorphology map of the basin. This unit appears darker than SFU in THEMIS IR-Day image. The valley networks surrounding the basin are seen to terminate into or before this unit. BDU has a fair distribution of craters with small diameter but lacks large-sized craters in comparison to SFU. It is covered with some bright deposits, which give it a rugged texture.

Division of the basin into two geological units helped in identification of the original and resurfaced regions on the basis of which age dating was performed. The differences in the morphology of the two units were examined in detail. Further, morphometric analysis of the features identified within the two units was carried out. It was observed that numerous craters are of different morphology and size ranging from a few hundred meters to tens of kilometers. Most of the large-sized craters in the BDU and SFU had partially degraded rims. SFU also had five craters with layered ejecta blankets. These craters have been named P, Q, R, S, and T and are shown individually in Figure 13.3-F, respectively.

A linear ridge was observed running through the center of the basin. This was found to be an extension of a previously reported dike network known to form in volatile-rich deposits. Several wrinkle ridges (Figure 13.3), distributed mostly in the central part of the basin (in the SFU), were seen. These are ~1 km wide and 50–100 m high from the surface and extend 35–45 km in the north-south direction. Low Strahler order values of the valley networks on the northern and western side as compared with those on the southern side indicate a less mature drainage system.

Studies propose that small tributaries that have distributed sources at the junction of drainage division are suggestive of a warmer and wetter Mars where an active hydrological cycle was responsible for constant recharging of the valley streams. Differences in the morphology and morphometry of the drainage channels on the three sides are suggestive of either different origin or multiple hydrological cycles. For instance, it is suggested that a V-shaped valley may modify to a U-shaped one when it reactivates, resulting into a widening of the valley in downward reaches. Hence, broadening of the valleys on the western side toward the termini may indicate another hydrological cycle in the basin at a later time.

Obliquity and pressure-driven migration of surface ice to lower latitudes favor a cold and dry Mars with episodic melting events. Thus, it is also possible that this difference in morphology of the valley networks may be the result of repeated freezing and thawing of ice, further hinting that this basin may have been hydrologically active for long periods. This also implies that a favorable climate existed wherein the temperatures may have exceeded the triple point, which led to melting of ice deposits within the region. Valley formation by erosion of hundreds of kilometers into the bedrock, high number of distributaries, and deep carving of the basin valleys as well as their resistance to erosional activities further confirm that these channels may be the result of the long and continuous action of water.

The accumulation of ice as well as its stability has been strongly correlated with obliquity of the planet (which changes in the cycle of $\sim 10^6$ years between 10° and 60°), the longitude of perihelion, and its eccentricity. It has been suggested that at an obliquity higher than 40°, the equator provides suitable conditions for snow and ice formation. Hence, it is possible that the basin experienced activities of liquid water and glaciations at different time periods.

Depth to diameter ratio in addition to CSFD plot were used in the present study to determine if any resurfacing processes have altered the craters over time. The shape of the CSFD curve can be used to understand the kind, duration, and intensity of the resurfacing event(s). Several geodynamic processes operating on Mars such as volcanic and glacial resurfacing, tectonic activity, erosion, and so on are responsible for alteration of the distribution of crater population. These events erase the existing crater population below a certain diameter and a new population accumulates presenting the time since resurfacing event. This resurfacing event appears as a "kink" in the cumulative fit CSFD plot. Absence of craters in the 16–50 , 8–13, and 2.7–4 km diameter range indicates that after the basin formed in the late Noachian, it experienced at least two major resurfacing events, first in the late Hesperian and then again in the middle Amazonian. Very low depth to diameter ratio of a large number of craters suggests that the craters have been filled since their formation and are thus more degraded.

The loss of the craters from the diameter ranging from 16 to 50 km suggests a resurfacing event, probably of high intensity, which completely erased these craters. The volcanic history of Mars is said to have begun around 4 Ga or probably earlier, which gradually decreased toward the late Hesperian. We suspect that the formation of Huygens basin, on the southern fringe of the basin, may have led to volcanic and/or tectonic resurfacing within the basin leading to abrupt cutoff falling in the range of the crater diameters below 50 km. The floors of large craters are more susceptible to modification due to fluvial sedimentation, volcanism, or other depositional processes as compared with the surrounding area. Wrinkle ridges have been known to form through contractual deformation and generally present in the volcanic plain. It is hypothesized that wrinkle ridges and the dike extension are remnants of a time when the basin was last geologically active, i.e. the late Amazonian (~440 Ma ago).

The morphological appearance of ejecta blankets and CSFD analysis of these craters do not suggest that the craters have witnessed uniform erosion. Pristine morphology of ejecta blankets of the craters and heavy erosion of ejecta blanket suggest that erosion does not correlate with the age of the craters. However, on the basis of the location of these craters, it is suggested that any volcanic resurfacing that took place within the basin was primarily concentrated within the central part of the basin. This may be because the lava was highly viscous and/or the event was less intense. Observation of single and double-layered ejecta craters suggests the presence of volatiles within the basin during the time of their impact.

Ejecta mobility (EM) provides an idea about the fluidity of the ejecta. High EM values may be attributed to low viscosity of the volatile material, high volatile content, or reduced friction between ejecta and the surface.

Desiccation of lake sediments, repeated freezing and thawing of the soil in peri glacial environments, thermal contraction of permafrost, and tectonic stress have been known to produce polygonal patterns on Earth similar to those reported on Mars. Although polygonal patterns have been found in abundance at the poles and higher latitudes, there have been fewer occurrences at lower latitudes. Meter-size polygons reported in the present study rule out tectonic stresses as the reason behind their formation. Although the average size of the polygons formed due to thermal contraction and desiccation is similar to those found here, the polygons observed in the current study are present inside the valleys and the terrain surrounding them. On age dating, it was found that this area belonged to the early Hesperian, a period during which Mars was hydrologically active. The presence of the polygons at a slightly higher elevation and toward the basin periphery reinforce the idea that the volcanic resurfacing, which was observed, was of low intensity and limited to the inner basin.

Furthermore, on comparison with previously reported polygonal patterns on Mars, it is found that Type 1 polygons resembled the third type of potential desiccated polygons reported in chloride-rich terrains in the vicinity of regions with phyllosilicate deposits. They have similar size and raised rim structure. Similarity of the polygons observed here to their counterparts reported elsewhere on Mars in addition to their location and size variation shows the mechanism behind their formation, and we interpret these polygons as plasma display panel (PDPs). Additionally, based on the mineralogy of the previous studies where PDPs have been identified, we suspect the presence of evaporites as well as clay minerals in the study area. The basin floor in a playa is composed of sediment clays, which makes the infiltration of water into the ground hard, which may have formed as tension developed in the sediment due to loss of water induced by changing climatic conditions such as high temperature and changing salinity of the depositing medium. When the water from the lake evaporates, it leaves behind minerals known as evaporites. These evaporites cement together and gradually increase in thickness.

Evaporites such as hydrous sulfates, halites, and carbonates have been reported in the Martian paleo lakes in addition to other more common deposits of clays or hydrated silica. Although silica minerals have been favored for holding astrobiological potential, Earth-based studies have been found to host endolithic bacteria abundance in several hyper arid conditions. It has been shown that not only hygroscopicity of some evaporite crystals (such as halites) helps in holding moisture, but the crystals also have the potential in mitigation of harmful UV radiation (e.g. gypsum). Thus, it has become imperative to study desiccation polygons and surrounding evaporite deposits in search for potential biosignatures. The present study shows that this inter-crater depression was hydrologically active in the past. The basin covers just about 20 000 square kilometers of the Martian surface. If all the hydrologically active inter-crater depressions are taken into consideration, they shall contribute toward a significant percentage of the area covered by the reservoirs of water. Thus, inter-crater depressions present a strong case for inclusion in water budgeting on Mars. Fluvial valley networks surrounding the basin on three sides have eroded possibly tens to hundreds of kilometers into the bedrock, suggesting the system was active for a long period. It is concluded that the basin was hydrologically active in the late Noachian wherein the lake formed. Pond formation of water may have altered the basaltic terrain or sediments could have been transported by the valley networks leading to deposition of phyllosilicate at the lake bottom, such as those reported in other paleo lake settings on Mars. The basin was hydrologically active in late Hesperian, wherein it eventually dried up and desiccation polygons were formed and evaporites were possibly deposited. Formation of crater "P" in the early-mid Amazonian period hints at the presence of water at the subsurface level, which may have later formed stable ice when the obliquity of Mars increased. Formation of the SLE craters "R" and "T" further support presence of volatiles (ice) during the middle Amazonian. As the obliquity changed again, volatiles evaporated followed by a less intensive volcanic resurfacing originating from the faults in central basin, in the late Amazonian, leaving behind smooth plains. From crater depth to diameter measurements, we find that the basin was around 1–2 km deep, which gradually filled with deposits (due to resurfacing events), reducing its depth to currently 300–500 m.The basin appears to be an ephemeral, endorheic playa which underwent several geodynamic processes. Based on the above interpretations, it is suggested that although the lake was last hydrologically active until the late Hesperian, the basin was geologically active until very recently, i.e. late Amazonian. The basin's distinct geological history and promising astro-paleo ecological potential make it a unique landing site for future exploration missions and setting up of permanent stations.

Further Reading

Ackiss, S.E., Seelos, K.D., and Buczkowski, D.L. (2013). Mineralogic mapping of Huygens crater, Mars: A transect of the highlands crust and Hellas basin rim. In *Large Meteorite Impacts and Planetary Evolution* 1737: 3095.

Ansan, V., Mangold, N., Masson, P. et al. (2008). Topography of valley networks on Mars from Mars Express High Resolution Stereo Camera digital elevation models. *Journal of Geophysical Research: Planets.* 113 (E7).

Bandfield, J.L., Hamilton, V.E., and Christensen, P.R. (2000). A global view of Martian surface compositions from MGS-TES. *Science.* 287 (5458): 1626–1630.

Barlow, N.G. (2005). A review of Martian impact crater ejecta structures and their implications for target properties. *Special Paper of the Geological Society of America.* 384: 433–442.

Barlow, N.G. and Bradley, T.L. (1990). Martian impact craters: correlations of ejecta and interior morphologies with diameter, latitude, and terrain. *Icarus.* 87 (1): 156–179.

Barlow, N.G. and Perez, C.B. (2003). Martian impact crater ejecta morphologies as indicators of the distribution of subsurface volatiles. *Journal of Geophysical Research: Planets.* 108 (E8).

Barlow, N.G., Boyce, J.M., Costard, F.M. et al. (2000). Standardizing the nomenclature of Martian impact crater ejecta morphologies. *Journal of Geophysical Research: Planets.* 105 (E11): 26733–26738. https://doi.org/10.1029/2000JE001258.

Bibring, J.P., Langevin, Y., Gendrin, A. et al. (2005). Mars surface diversity as revealed by the OMEGA/Mars Express observations. *Science.* 307 (5715): 1576–1581.

Bishop, J.L., Tirsch, D., Tornabene, L.L. et al. (2013). Mineralogy and morphology of geologic units at Libya Montes, Mars: ancient aqueously derived outcrops, mafic flows, fluvial features, and impacts. *Journal of Geophysical Research: Planets.* 118 (3): 487–513.

Cabrol, N.A. and Grin, E.A. (1999). Distribution, classification, and ages of Martian impact crater lakes. *Icarus.* 142 (1): 160–172.

Cabrol, N.A. and Grin, E.A. (2001). The evolution of lacustrine environments on Mars: Is Mars only hydrologically dormant? *Icarus* 149 (2): 291–328.

Carr, M.H. and Head, J.W. III, (2010). Geologic history of Mars. *Earth and Planetary Science Letters.* 294 (3-4): 185–203.

Craddock, R.A. and Howard, A.D. (2002). The case for rainfall on a warm, wet early Mars. *Journal of Geophysical Research: Planets.* 107 (E11): 21–21.

Davila, A.F., Gómez-Silva, B., de Los Rios, A. et al. (2008). Facilitation of endolithic microbial survival in the hyperarid core of the Atacama Desert by mineral deliquescence. *Journal of Geophysical Research: Biogeosciences.* 113 (G1): 1–9.

Dong, H., Rech, J.A., Jiang, H. et al. (2007). Endolithic cyanobacteria in soil gypsum: occurrences in Atacama (Chile), Mojave (United States), and Al-Jafr Basin (Jordan) Deserts. *Journal of Geophysical Research: Biogeosciences.* 112 (G2): 1–11.

Edwards, C.S., Nowicki, K.J., Christensen, P.R. et al. (2011). Mosaicking of global planetary image datasets: 1. Techniques and data processing for Thermal Emission Imaging System (THEMIS) multi-spectral data. *Journal of Geophysical Research: Planets.* 116 (E10): 1–13.

Ehlmann, B.L., Mustard, J.F., Murchie, S.L. et al. (2008). Orbital identification of carbonate-bearing rocks on Mars. *Science.* 322 (5909): 1828–1832.

El-Maarry, M.R., Dohm, J.M., Michael, G. et al. (2013). Morphology and evolution of the ejecta of Hale crater in Argyre basin, Mars: results from high resolution mapping. *Icarus.* 226 (1): 905–922.

El-Maarry, M.R., Pommerol, A., and Thomas, N. (2013). Analysis of polygonal cracking patterns in chloride-bearing terrains on Mars: Indicators of ancient playa settings. *Journal of Geophysical Research: Planets.* 118 (11): 2263–2278.

El-Maarry, M.R., Watters, W., McKeown, N.K. et al. (2014). Potential desiccation cracks on Mars: a synthesis from modeling, analogue-field studies, and global observations. *Icarus.* 241: 248–268.

Fassett, C.I. and Head, J.W. III, (2008a). The timing of Martian valley network activity: constraints from buffered crater counting. *Icarus.* 195 (1): 61–89.

Fassett, C.I. and Head, J.W. III, (2008b). Valley network-fed, open-basin lakes on Mars: distribution and implications for Noachian surface and subsurface hydrology. *Icarus.* 198 (1): 37–56.

Fawdon, P., Gupta, S., Davis, J.M. et al. (2018). The Hypanis Valles delta: the last highstand of a sea on early Mars? *Earth and Planetary Science Letters.* 500: 225–241.

Fergason, R., Hare, T., and Laura, J. (2017). *HRSC and MOLA blended digital elevation model at 200m v2, Astrogeology PDS Annex*. Geological Survey: U.S http://bit.ly/HRSC_MOLA_Blend_v0 (accessed 11 January 2017).

Garvin, J.B., Sakimoto, S.E., Frawley, J.J., and Schnetzler, C. (2000). North polar region crater forms on Mars: geometric characteristics from the Mars Orbiter Laser Altimeter. *Icarus.* 144 (2): 329–352.

Gillette, D.A., Adams, J., Endo, A. et al. (1980). Threshold velocities for input of soil particles into the air by desert soils. *Journal of Geophysical Research: Oceans.* 85 (C10): 5621–5630.

Goudge, T.A., Aureli, K.L., Head, J.W. et al. (2015). Classification and analysis of candidate impact crater-hosted closed-basin lakes on Mars. *Icarus.* 260: 346–367.

Goudge, T.A., Head, J.W., Mustard, J.F., and Fassett, C.I. (2012). An analysis of open-basin lake deposits on Mars: evidence for the nature of associated lacustrine deposits and post-lacustrine modification processes. *Icarus.* 219 (1): 211–229.

Greeley, R. and Spudis, P.D. (1981). Volcanism on mars. *Reviews of Geophysics.* 19 (1): 13–41.

Gulick, V.C. (2001). Origin of the valley networks on Mars: a hydrological perspective. *Geomorphology.* 37 (3-4): 241–268.

Head, J.W., Wilson, L., Dickson, J., and Neukum, G. (2006). The Huygens-Hellas giant dike system on Mars: implications for Late Noachian–Early Hesperian volcanic resurfacing and climatic evolution. *Geology.* 34 (4): 285–288.

Howard, A.D., Moore, J.M., and Irwin, R.P. III, (2005). An intense terminal epoch of widespread fluvial activity on early Mars: 1. Valley network incision and associated deposits. *Journal of Geophysical Research: Planets.* 110 (E12).

Hynek, B.M., Beach, M., and Hoke, M.R. (2010). Updated global map of Martian valley networks and implications for climate and hydrologic processes. *Journal of Geophysical Research: Planets.* 115 (E9).

Irwin, R.P. III, Lewis, K.W., Howard, A.D., and Grant, J.A. (2015). Paleohydrology of Eberswalde crater, Mars. *Geomorphology.* 240: 83–101.

Irwin, R.P. and Howard, A.D. (2002). Drainage basin evolution in Noachian Terra Cimmeria. *Mars. Journal of Geophysical Research: Planets.* 107 (E7): 10–23.

Ivanov, B.A. (2001). Mars/Moon cratering rate ratio estimates. *Space Science Reviews.* 96 (1-4): 87–104.

Kerber, L., Dickson, J.L., Head, J.W., and Grosfils, E.B. (2017). Polygonal ridge networks on Mars: diversity of morphologies and the special case of the Eastern Medusae Fossae Formation. *Icarus.* 281: 200–219.

Kite, E.S. (2019). Geologic constraints on early Mars climate. *Space Science Reviews.* 215 (1): 10.

Kneissl, T., van Gasselt, S., and Neukum, G. (2011). Map-projection-independent crater size-frequency determination in GIS environments—new software tool for ArcGIS. *Planetary and Space Science.* 59 (11–12): 1243–1254.

Lamb, M.P., Dietrich, W.E., Aciego, S.M. et al. (2008). Formation of Box Canyon, Idaho, by megaflood: Implications for seepage erosion on Earth and Mars. *Science.* 320 (5879): 1067–1070.

Laskar, J., Correia, A.C.M., Gastineau, M. et al. (2004). Long term evolution and chaotic diffusion of the insolation quantities of Mars. *Icarus.* 170 (2): 343–364.

Malin, M.C., Bell, J.F., Cantor, B.A. et al. (2007). Context camera investigation on board the Mars Reconnaissance Orbiter. *Journal of Geophysical Research: Planets.* 112 (E5): 1–25.

Mansfield, M., Kite, E.S., and Mischna, M.A. (2018). Effect of Mars atmospheric loss on snow melt potential in a 3.5-gyr Mars climate evolution model. *Journal of Geophysical Research: Planets.* 123 (4): 794–806.

Matsubara, Y., Howard, A.D., and Irwin, R.P. III, (2018). Constraints on the Noachian paleoclimate of the Martian highlands from landscape evolution modeling. *Journal of Geophysical Research: Planets.* 123 (11): 2958–2979.

McEwen, A.S., Eliason, E.M., Bergstrom, J.W. et al. (2007). Mars Reconnaissance Orbiter's high resolution imaging science experiment (HiRISE). *Journal of Geophysical Research: Planets.* 112 (E5): 1–40.

McEwen, A.S., Preblich, B.S., Turtle, E.P. et al. (2005). The rayed crater Zunil and interpretations of small impact craters on Mars. *Icarus.* 176 (2): 351–381.

McMahon, S., Bosak, T., Grotzinger, J.P. et al. (2018). A field guide to finding fossils on Mars. *Journal of Geophysical Research: Planets.* 123.

Melosh, H.J. (1989). *Impact cratering: A geologic process. Research supported by NASA*, Oxford Monographs on Geology and Geophysics No. 11, 253pp. New York: Oxford University Press.

Michael, G.G. and Neukum, G. (2010). Planetary surface dating from crater size–frequency distribution measurements: partial resurfacing events and statistical age uncertainty. *Earth and Planetary Science Letters.* 294 (3-4): 223–229.

Mischna, M.A., Richardson, M.I., Wilson, R.J., and McCleese, D.J. (2003). On the orbital forcing of Martian water and CO2 cycles: a general circulation model study with simplified volatile schemes. *Journal of Geophysical Research: Planets.* 108 (E6).

Montgomery, D.R. (2002). Valley formation by fluvial and glacial erosion. *Geology.* 30 (11): 1047–1050.

Mouginis-Mark, P. (1979). Martian fluidized crater morphology: variations with crater size, latitude, altitude, and target material. *Journal of Geophysical Research: Solid Earth.* 84 (B14): 8011–8022.

Mukherjee, S., Singh, D., Singh, P., and Roy, N. (2020). Morphological and morphometric analysis of a topographic depression near Huygens basin, Mars: identification of a putative endorheic playa. *Geomorphology* 351: 106912.

Osterloo, M.M., Anderson, F.S., Hamilton, V.E., and Hynek, B.M. (2010). Geologic context of proposed chloride-bearing materials on Mars. *Journal of Geophysical Research: Planets.* 115 (E10).

Palucis, M.C., Dietrich, W.E., Williams, R.M. et al. (2016). Sequence and relative timing of large lakes in Gale crater (Mars) after the formation of Mount Sharp. *Journal of Geophysical Research: Planets.* 121 (3): 472–496.

Pechmann, J.C. (1980). The origin of polygonal troughs on the northern plains of Mars. *Icarus.* 42 (2): 185–210.

Platz, T., Michael, G., Tanaka, K.L. et al. (2013). Crater-based dating of geological units on Mars: methods and application for the new global geological map. *Icarus.* 225 (1): 806–827.

Plescia, J.B. and Golombek, M.P. (1986). Origin of planetary wrinkle ridges based on the study of terrestrial analogs. *Geological Society of America Bulletin.* 97 (11): 1289–1299.

Plummer, P.S. and Gostin, V.A. (1981). Shrinkage cracks: desiccation or synaeresis? *Journal of Sedimentary Research.* 51 (4).

Robbins, S.J. and Hynek, B.M. (2012a). A new global database of Mars impact craters ≥ 1 km: 2. Global crater properties and regional variations of the simple-to-complex transition diameter. *Journal of Geophysical Research: Planets.* 117 (E6): https://doi.org/10.1029/2011JE003967.

Robbins, S.J. and Hynek, B.M. (2012b). A new global database of Mars impact craters ≥ 1 km: 1. Database creation, properties, and parameters. *Journal of Geophysical Research: Planets.* 117 (E6): https://doi.org/10.1029/2011JE003966.

Rodríguez-Rodríguez, M., Moral, F., Benavente, J., and Beltrán, M. (2010). Developing hydrological indices in semi-arid playa-lakes by analyzing their main morphometric, climatic and hydrochemical characteristics. *Journal of Arid Environments.* 74 (11): 1478–1486.

Salese, F., Pondrelli, M., Neeseman, A. et al. (2019). Geological evidence of planet-wide groundwater system on Mars. *Journal of Geophysical Research: Planets.* 124 (2): 374–395.

Schultz, R.A. (2000). Localization of bedding plane slip and backthrust faults above blind thrust faults: keys to wrinkle ridge structure. *Journal of Geophysical Research: Planets.* 105 (E5): 12035–12052.

Strahler, A.N. (1957). Quantitative analysis of watershed geomorphology. *Eos, Transactions American Geophysical Union.* 38 (6): 913–920.

Watters, T.R. (1991). Origin of periodically spaced wrinkle ridges on the Tharsis Plateau of Mars. *Journal of Geophysical Research: Planets.* 96 (E1): 15599–15616.

Werner, S.C. (2008). The early Martian evolution—Constraints from basin formation ages. *Icarus.* 195 (1): 45–60.

Wierzchos, J., Cámara, B., De Los Rios, A. et al. (2011). Microbial colonization of Ca-sulfate crusts in the hyperarid core of the Atacama Desert: implications for the search for life on Mars. *Geobiology.* 9 (1): 44–60.

Williams, D.A., Greeley, R., Manfredi, L. et al. (2010). The Circum-Hellas volcanic province, Mars: assessment of wrinkle-ridged plains. *Earth and Planetary Science Letters.* 294 (3-4): 492–505.

Williams, R.M. and Phillips, R.J. (2001). Morphometric measurements of Martian valley networks from Mars Orbiter Laser Altimeter (MOLA) data. *Journal of Geophysical Research: Planets.* 106 (E10): 23737–23751.

Wordsworth, R.D., Kerber, L., Pierrehumbert, R.T. et al. (2015). Comparison of "warm and wet" and "cold and icy" scenarios for early Mars in a 3-D climate model. *Journal of Geophysical Research: Planets.* 120 (6): 1201–1219.

Wordsworth, R., Forget, F., Millour, E. et al. (2013). Global modelling of the early Martian climate under a denser CO2 atmosphere: water cycle and ice evolution. *Icarus.* 222 (1): 1–19.

Wray, J.J., Ehlmann, B.L., Squyres, S.W. et al. (2008). Compositional stratigraphy of clay-bearing layered deposits at MawrthVallis. *Mars. Geophysical Research Letters.* 35 (12): 1–6.

14

Future of Extraterrestrial Remote Sensing

14.1 Introduction

Environmental perturbations and sudden episodic events can be addressed by the understanding of extraterrestrial changes. These changes are monitored by terrestrial as well as extraterrestrial satellites. Terrestrial satellites need to be strengthened in accordance with extraterrestrial findings between the Sun and the Earth. Slow changes on Earth and environmental processes like tectonic movements can be monitored by very high-resolution low orbit or high orbit fast-moving satellites, while sudden cyclones, tornadoes, global warming or cooling, and glacial melting are processes of sudden change which appear to have impact from outer space. Any changes on the Earth and environment that is named a natural disaster should be given maximum importance in extraterrestrial satellite investigation. However, it cannot be ruled out to plan the deep space satellite, which will monitor the galactic influence on the Sun–Earth and all living–nonliving things on the Earth. It is established that the whole universe is expanding continuously and the magnetic field is affected due to this expansion. Sensors are required to measure the relative magnetic field in between different celestial bodies. Orientation of differently charged molecules of all planetary materials needed to be mapped. Like the universe, the human body also has magnetic poles, which are bounded by the invisible string of a magnetic field, which changes with time and place. Sun distant stars and planets all have individual magnetic fields, and they influence living systems including the human body. It is essential to develop sensors to infer climate change and its effect on natural and human-made health stressors. Between Sun and the Earth, the magnetic field changes, which can have a profound effect on the human immune system and health. In addition, the changes in hydrological cycles, pollution potentiality, and other environmental stressors can increase possibilities for disease, especially in vulnerable populations like South Asia. As inferred by solar experts, we cannot rule out the possibility of sun-induced warming of the climate, which may increase exposure to allergens and disrupt antigen-specific tolerance, leading to a rise in immunologic disease. Dengue, Malaria, other water-generated illness, and emerging pathogens, such as SARS-CoV-2, can cause epidemics that result in profound disruptions in global health security. Food security can be challenged by improper diet patterns required due to climate change, which can affect increasing susceptibility to disease, especially among children. Sensors should be developed to address not only these issues, but the micro biome surge mapping in air and water will be of prime importance. A new investigation monitoring pixel-based changes from satellites can be initiated to monitor the release of hydrogen peroxide from earthquake-prone areas to be a mitigating solution in the future. Delhi and its surrounding areas of India have shown influence of neotectonic activities, which has the potential to release hydrogen peroxide from the stressed interstices of quartzite rocks. Monitoring of stress within bedrocks along active fault

Extraterrestrial Remote Sensing and Climate Change, First Edition. Saumitra Mukherjee.

zones has been under observation for detecting premonitory earthquake signals. Quartzite, basalt, and other metamorphic and igneous rocks can produce hydrogen peroxide under stressed conditions. These stress-activated defects form highly reactive oxygen species (ROS), which produce electric currents. Where the rock is developing stress, the ROS can combine with groundwater to produce hydrogen peroxide, giving an indirect measure of the underlying rock stress along active fault planes within hard rock aquifers and deep thermal springs. Aquifers under stress can influence both the quality and quantity of groundwater. Frequent change in groundwater levels may lead to mixing of water between different aquifers, influx of water from surrounding areas, changes in dissolved gas concentrations, and mineral dissolution in the fissures of the rocks through infiltration of pollutants from soil or ground surface. It has been recorded that the rise in microtremor in the months of December 2020 and January 2021 in Delhi in particular and in India and surrounding areas of India in general has reduced COVID-19 cases dramatically (https://health.economictimes.indiatimes.com/news/diagnostics/why-indias-coronavirus-cases-have-fallen-so-sharply/80704609). However, COVID-19 cases are on the rise across India and the world in March–April–May 2021, but it shows a decrease in cases in Assam, Mizoram, Nagaland, and other northeastern parts of India where recent tectonic microtremor was recorded (https://www.newindianexpress.com/nation/2021/feb/17/assam-reports-fivenew-covid-19-cases-lowest-since-april-2020-2264970.html https://www.volcanodiscovery.com/region/7971/earthquakes/assam.html

The hypothesis is strongly supported by the presence of pulverized lunar regolith formed by tectonic activity and possible release of high concentration of hydrogen peroxide from the tectonically active areas. Reactivation of an active fault line potential for emergence of earthquake in Delhi, India has been reported in 2001 as Asola Bahadurgarh fault. Similarly, other parts of India have shown the emergence of new fault lines. The correlation of the generation of new fault lines and its release of hydrogen peroxide to mitigate the deadly pandemics can be a new correlation and can be used across the world to mitigate future pandemics by generating microtremor or artificial earthquakes to release the hydrogen peroxide from the land of the affected terrain.

14.2 High-resolution Extraterrestrial Investigation

It will be mandatory to infer the terrestrial and extraterrestrial resources by very high-resolution satellite data. There are some new satellite constellations planned to be launched by France, namely Pléiades Neo 4. Pléiades Neo 4 was released from the launch vehicle, very close to its final 620-km Sun-synchronous polar orbit, which it will reach within a few days. The satellite will be phased 180° with Pléiades Neo 3 on the same orbit to start forming a constellation. This will enable daily imaging of any place on Earth at 30 cm native resolution, and between two and four times a day when the four-satellite constellation is complete. "Pléiades Neo will offer a truly best-in-class capability to our customers and will strongly enhance our position in the very high-resolution market" said François Lombard, Head of Intelligence at Airbus Defence and Space. "The first images from Pléiades Neo 3 are outstanding and confirm that we took the right decision in terms of design and performance to address the increasingly demanding requirements of the geospatial sector." Comprising four identical satellites, the 100% Airbus manufactured, owned, and operated Pléiades Neo constellation offers a native resolution of 30 cm with an imaging swath of 14 km, the widest in its category. Thanks to their unmatched agility, the constellation will be able to cover the entire Earth landmass five times per year. The new satellites will work hand in hand with the existing Pléiades satellites and the rest of the Airbus dozen-strong Earth observation satellite fleet. The highly innovative design of the Pléiades Neo spacecraft is equipped with the next-generation silicon carbide optical instrument, building on the technology that Airbus first pioneered in the early 2000s. The Pléiades Neo constellation will also benefit from laser optical and Ka-band links with the Airbus Space Data Highway (EDRS) geostationary satellites to enable urgent acquisitions less than 40 minutes after tasking, to swiftly respond to the most critical situations. Besides the understanding of the life and life support system of the Earth and other planets, the future satellites will focus on Asteroid mapping for its resources and possible impact on the Earth.

Satellite OSIRIS-REx was launched by NASA, USA, on 20 October 2020. The spacecraft was placed at 320 million kilometers from the Earth and grabbed dust and rocks from a sample site called Nightingale (Figure 14.1). The samples collected are approximately weighing 60 g, which will reach the Earth in September 2023 from a potentially hazardous Bennu asteroid belt. Bennu is a remnant of large carbon-rich asteroid complex, which was formed in the main Asteroid Belt between Mars and Jupiter and since then it has migrated toward the Earth (Figure 14.2).

A 60-year-old mystery regarding the source of some energetic and potentially damaging particles in the Earth's radiation belts has been solved, using data from a student-operated shoebox-sized satellite. The results indicate energetic electrons in Earth's inner radiation belt – primarily near its inner edge – are created by cosmic rays born from explosions of supernovas. Earth's radiation belts, known as the Van Allen belts, are layers of energetic particles held in place by the planet's magnetic field.

When the magnetic field of the planets changes between the Sun–Planet environment, changes in the environment can be expected. It is pertinent to think that the movement of each and every environmental component and cycle of the Earth depends on the magnetic field fluctuation of the Earth, Sun, and in between Sun and the Earth.

A Van Allen radiation belt is a zone of energetic charged particles, most of which originate from the solar wind, that are captured by and held around a planet by that planet's magnetosphere. Earth has two such belts, and sometimes others may be temporarily created (Figure 14.3).

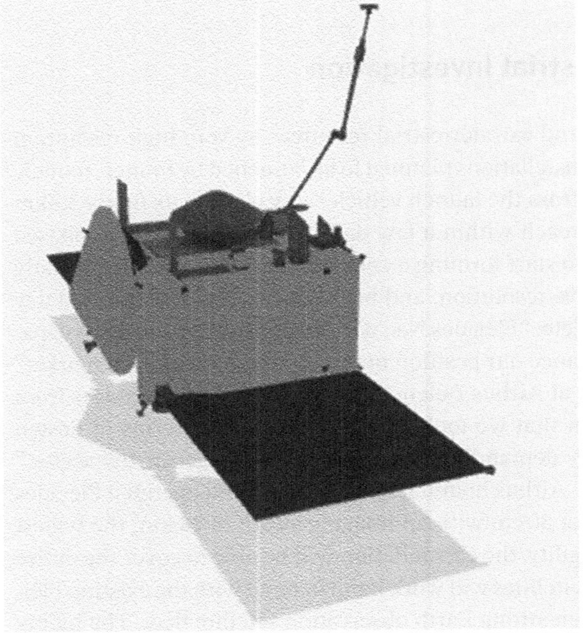

Figure 14.1 Satellite OSIRIS-REx dedicated for mapping Asteroids.

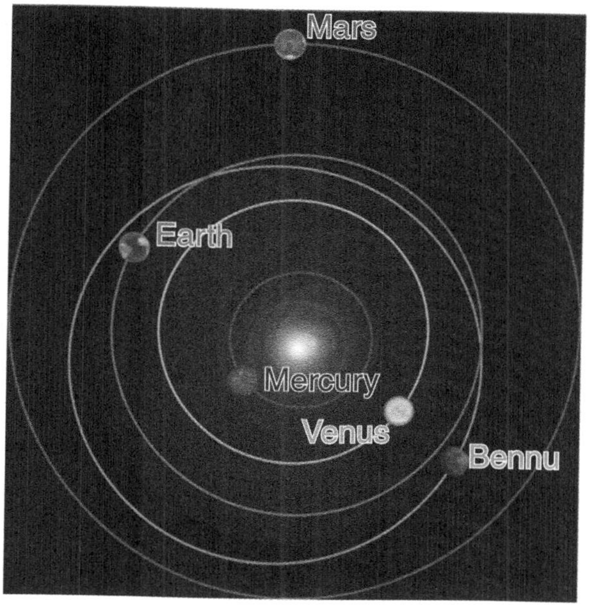

Figure 14.2 Position of Bennu Asteroid Belt between Mars and the Earth. Source: Orbit of Asteroid Bennu.

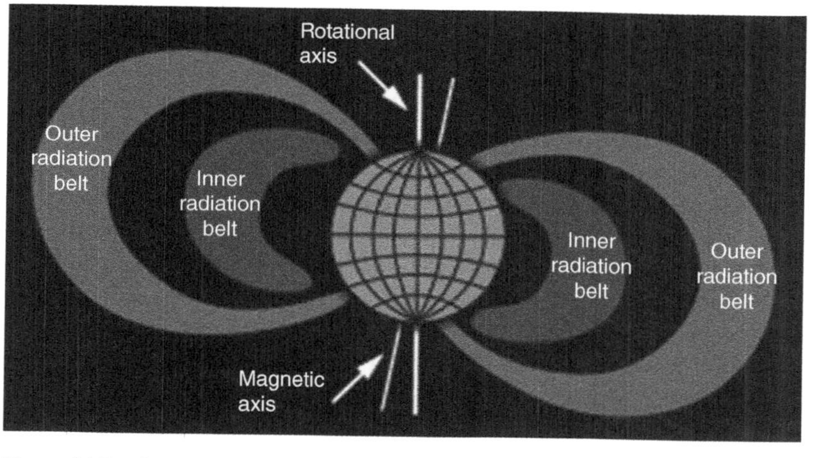

Figure 14.3 Schematic diagram of Van Allen belts showing impact of solar radiation on the Earth. Source: NASA.

During a process called "cosmic ray albedo neutron decay" (CRAND), cosmic rays entering Earth's atmosphere collide with neutral atoms, creating a "splash," which produces charged particles, including electrons that become trapped by Earth's magnetic fields. The findings have implications for understanding and better forecasting the arrival of energetic electrons in near-Earth space, which can damage satellites and threaten the health of astronauts.

14.3 Ionosphere Monitoring Satellite

Ukraine National Space Programme has developed Ionosat-Micro satellite and France has developed DEMETER ionosphere measuring satellite to monitor seismic and volcanic precursor.

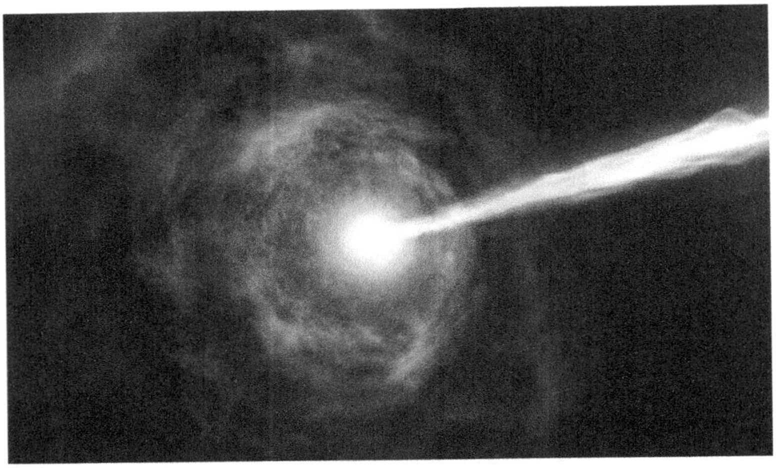

Figure 14.4 Firce Gama Ray Burst in 2019. Source: Credits: NASA, ESA and M. Kornmesser.

The main objective of the mission is to monitor dynamic processes in the ionosphere through concerted space and ground-based measurements to find the relationship of ionospheric disturbances with processes on the Sun, in the magnetosphere, the atmosphere, and the Earth's inner shells. These satellites are Sun synchronous and placed at 600–700 km above the Earth. It is not clear about the controlling factor of the Sun–Earth interface. Supporting information on the change of the interface may lie with the observations from NASA's Hubble Space Telescope. It has the potential to investigate the nature of the powerful gamma-ray burst GRB by studying its environment. Figure 14.4 shows that gamma-ray bursts of 2019 are the most powerful explosions in the universe. GRB emit most of their energy in gamma rays, light that is much more energetic than the visible light we can see with our eyes. Hubble's observations suggest that this particular burst displayed such powerful emission due to the collapsing star that was sitting in a very dense environment, right in the middle of a bright galaxy 5 billion light years away. In 2019, the gamma-ray burst was unprecedented, followed by the outbreak of COVID-19. It is also important to note that the solar maximum was due in 2015–2016, but it was a very mediocre event. The extraterrestrial process has changed, which includes the controlled activities of the Sun that inspired the thermosphere–ionosphere–atmosphere–hydrosphere–biosphere perturbation. In a nutshell, it can be stated that climate change has been governed by the series of events of GRB-low solar activity and change in the environment. It is essential to have more satellite-based GRB observatories for accurate prediction of climate change.

Besides Spacecraft (SC) "Microsat-M" (Figure 14.5) service systems is also available, and it provides following the facilities:

- GPS determination of SC positioning with error about $\sim$ 20 m
- SC attitude control within 5°
- SC attitude determination with the error no more than $\sim$0.2°.

Scientific payload. The scientific payload "Ionosat-Micro" is intended for registration of the following formations in the ionosphere (Table 14.1):

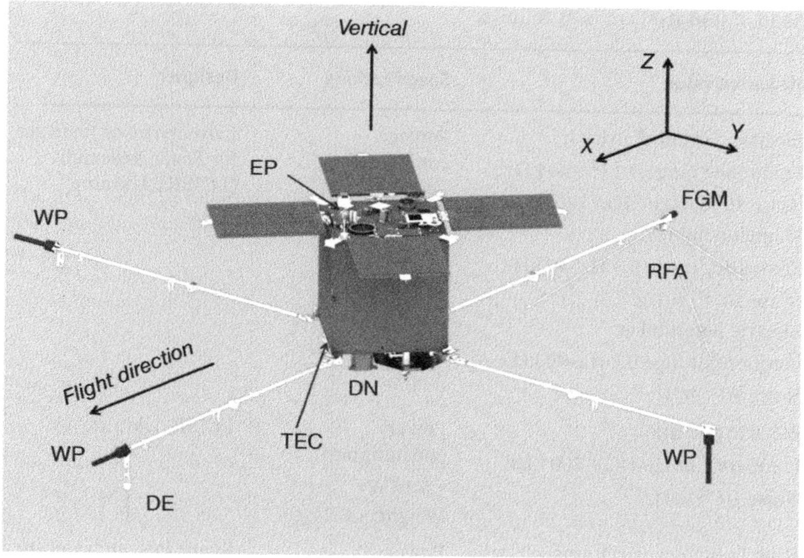

Figure 14.5 Spacecraft "Microsat" with booms.

- Space temporal structure and global distribution of non-homogeneous atmosphere;
- Global structure and dynamics of magnetosphere–ionosphere current system and geomagnetic field;
- Electromagnetic wave structures and turbulences in the ionosphere.

To carry out such research, the scientific payload is to be provided with measurements of the following parameters:

- Neutral gas and plasma parameters – concentrations, temperatures;
- Vectors of DC-ELF-VLF electromagnetic field and ELF-VLF plasma current fluctuations (waveforms);
- Total electron content (TEC);
- Spectrum of ionospheric plasma waves.

The toolkit of scientific instrumentation includes the sensors of electromagnetic fields and plasma currents in the ionosphere EP, WP, and FGM (developer Lviv Center of Institute of Space Research NASU & SSAU), sensors of neutral gas and plasma parameters DN and DE (developer Institute for Technical Mechanics NASU & SSAU), radiofrequency analyzer of plasma wave spectrum RFA (Space Research Center of Polish Academy of Science).

In 2023, India will launch Aditya satellite for monitoring extraterrestrial phenomena. Thus, the enhanced Aditya project will enable a comprehensive understanding of the dynamic processes of the sun and address some of the outstanding problems of heliophysical parameters to correlate climate change.

To understand the impact of extraterrestrial changes on the planets, specifically the pole shift of the Earth needed to compare with the Moon of the Earth. Chandrayaan-1 Data of Indian Space Research Organization satellite has initiated the possibility to infer tectonic activity of the South Polar Region of the Moon. Microwave sensor (Mini-SAR) data from Chandrayaan-1 satellite was investigated using ENVI software. It is being reported that the geological processes on the Moon have found various tectonic features suggesting similar geological activity occurring on the Earth. Further, it has been found out that cosmic rays and solar wind flux variability coupled with the

Table 14.1 Scientific payload of Inosat-Micro is as follows.

Sensor	Measured value	Specifications	Designer
1. Wave probes WP (3 pieces)	Electric current density J: Frequency range 0.1 Hz–40 kHz Noise 10^{-12} A\cm^{2}Hz$^{1/2}$ Magnetic field B: Frequency range 0.1 Hz–40 kHz Noise 10^{-14} T/Hz$^{1/2}$ Electric potential φ: Frequency range 0.1 Hz–40 kHz Noise 10^{-6} V/Hz$^{1/2}$	Power consumption: <0.25 W Weight: 0.225 kg	Lviv Centre of Institute for Space Research (LCISR), Ukraine
2. Electric probe EP	Electric potential: Frequency range DC – 200 kHz Noise 10^{-6} V/Hz$^{1/2}$	Power consumption: <0.15 W Weight: <0.2 kg	LCISR, Ukraine
3. Radiofrequency analyzer RFA	High frequency variations, electric component Frequency range 0.1–15 MHz	Power consumption: <5 W Weight: ≤4 kg (including antenna)	Space Research Centre, Poland
4. Sensor of neutral and charged particles DN-DE	Density and temperature neutral particle N_n: 10^5–10^{12} cm^{-3} charged particle N_i: 10^3–10^{11} cm^{-3} Electron temperature 0.1–1.5 keV	Power consumption: ≤2 W Weight: ≤3 kg	Institute of Technical Mechanics, Ukraine
5. DC flux-gate magnetometer FGM	Frequency range DC – 1 Hz Resolution 0.01 nT	Power consumption: <0.3 W. Weight: ≤0.3 kg.	LCISR, Ukraine
6. TEC meter	Frequency L1 = 1217–1265, L2 = 1565–1615 MHz, 20 channels	Power consumption: 2.4 W, total Weight: ≤1.34 kg	IZMIRAN, Russia
7. DPU	Input information rate, Mb/c, 100 Output information rate, Mb/c, 64 Onboard memory, up to 28 GB	Power consumption: <4 W Weight: ≤1.5 kg	LCISR, Ukraine

interior tectonic activity and meteorite impact can be responsible for tectonic activity on the Moon. This finding is supported by the manifestation of tectonic activities on the Moon in the form of faults and graben. Previous studies investigating lunar tectonic activity have used optical data primarily to study lunar tectonics. Present work confirms that partially and completely shadowed crater interiors present in the polar regions can also contain unique features indicative of tectonic activity (Figure 14.6). The study therefore focuses on the identification of features indicating the presence of faults/lineaments/fractures on the lunar topography (Figure 14.7).

Mini-SAR imaging radar onboard Chandrayaan-1 mission was the first mono-static lunar orbiting synthetic aperture radar. Mini-SAR data is primarily used to identify signatures suggesting the

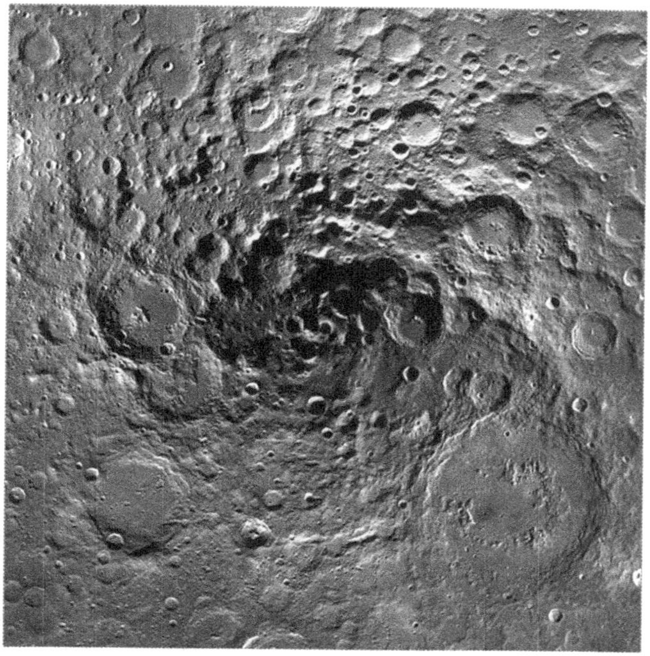

Figure 14.6 South Polar Region of the Moon (Schrödinger basin). Source: NASA.

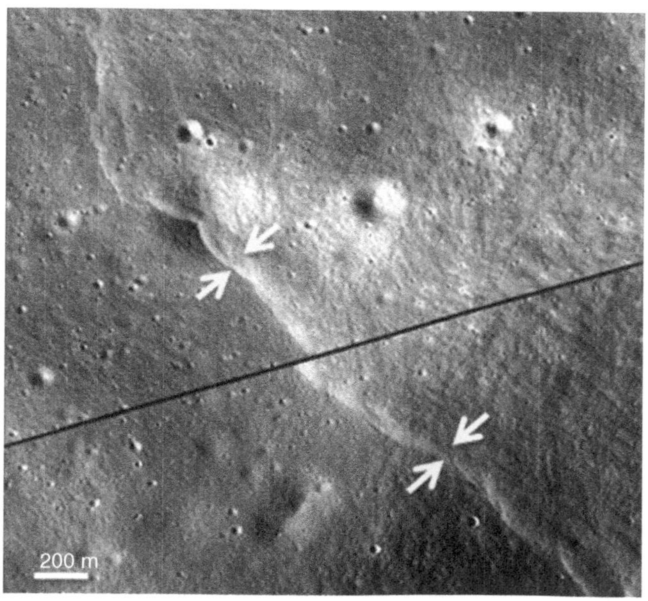

Figure 14.7 Scarp lying between roughly 71.29°S, 100.13°W, and 71.34°S, 103.01°W: Scarp length ~20 km in the South Pole of the Moon.

possibility of the presence of water ice within the completely shadowed regions of the lunar poles. Further, this data has been used to view various other geological features such as melt flows, crater ejecta blankets, secondary craters, and so on. Therefore, identification of distinct morphological features in shadowed as well as illuminated regions of lunar surfaces can be easily done using Mini-SAR data. The SAR technology can also penetrate the surface and therefore is an effective tool to look for hidden fault lines in the shadowed portions on the lunar regolith. Alternatively, other distinct surface and subsurface geological features have also been identified which can further help to study the process of impact cratering as well as understand the formation of secondary features developed post impact.

The structural geological information on the surface of the Moon suggests present and past tectonic activities on this satellite. These structural features are comparable with the features of the Earth. Presence of solid masses of water has also been inferred within the volcanic craters.

This investigation is useful in view of future manned and unmanned ventures and suitable areas for the landing and scientific investigation of the different planets of the solar system.

It is well known that global risks are increasingly complex, uncertain, systemic, and dynamic. The recent pandemic and catastrophic changes have drawn attention to the critical importance of global risks around the world. To tackle global risks effectively, we will need a stronger understanding of the likelihood, impact, and linkages between wide ranges of risks. It is recommended to explore more safe land areas on the Earth's surface, which can be colonized to tackle the increasing population and its demand for land, food, and water resources.

The space environment is harsh not only on humans and on other living organisms, but also on instruments. Damage from solar energetic particles and cosmic rays can slowly degrade performance of an instrument (Figure 14.8). Fortunately, there are ways to characterize and correct for this degradation (Figure 14.9).

The sensors of extraterrestrial satellites need to be updated very quickly. It is estimated that sensitivity of the sensors must be improved by launching new satellites or it may be repaired onsite.

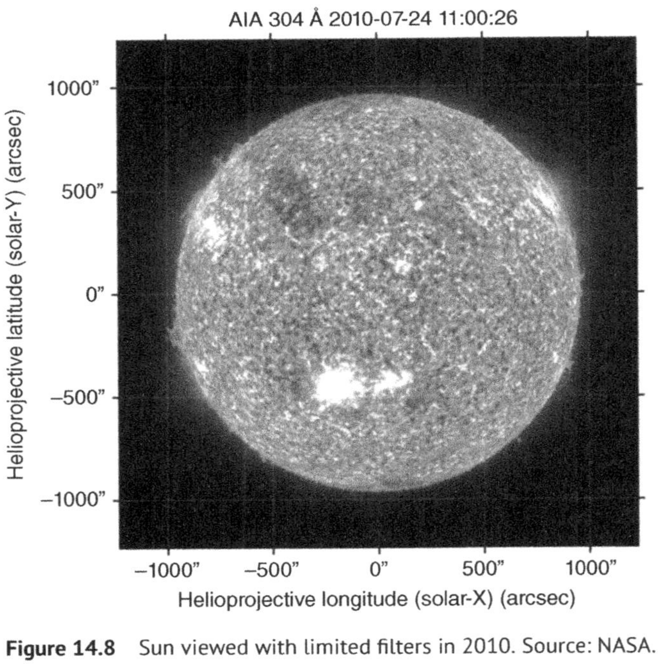

Figure 14.8 Sun viewed with limited filters in 2010. Source: NASA.

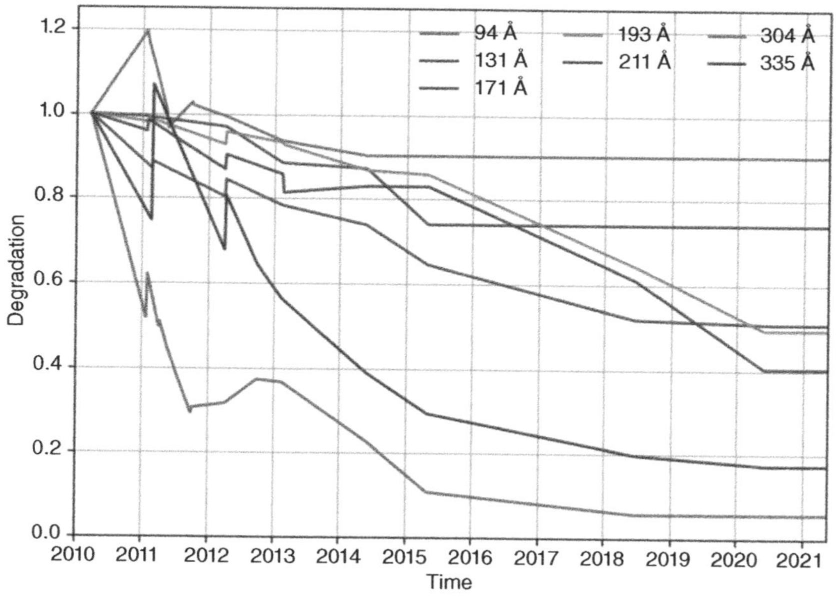

Figure 14.9 Degradation of instrumentation sensitivity with time for the AAI instruments on SDO. Source: Courtesy NASA.

Besides the impact of the Sun on the space weather, some natural as well as manmade activities like a nuclear explosion may perturb the magnetic field between the Sun and the Earth. A nuclear explosion in the near-Earth space environment can produce an electromagnetic pulse (EMP) at the Earth's surface or below. Geoelectric field can be induced very quickly, with potential grid failure of high-tension power lines. It can disrupt mobile phone connectivity and GPS devices. The geo-electric field variation may have impact on the aquifer solubility and conductivity of rock structures. Future satellites should be equipped with sensors to monitor the electric field variation around the Earth, and may develop a warning system or have a mitigation solution.

Besides these, upcoming solar system exploration missions will search for extraterrestrial (ET) life, but extraterrestrial life may not be like Earth life. A new mass spectrometry analysis technique may allow for process-based ways to find ET life that is compositionally alien. The investigation of extraterrestrial life in the Martian soil has already been shown to be a possibility. Such life may be subtly or profoundly different from Earth life, and methods based on detecting particular molecules as biosignatures may not apply to life with a different evolutionary history. A new study by a joint Japan/US-based team, led by researchers at the Earth-Life Science Institute (ELSI) at the Tokyo Institute of Technology, has developed a machine learning technique which assesses complex organic mixtures using mass spectrometry to reliably classify them as biological or a biogeochemical change. Exciting missions such as NASA's *Perseverance* rover will look for life this year on Mars; NASA's *Europa Clipper*, launching in 2024, will try to sample ice ejected from Jupiter's moon Europa, and its *Dragonfly* mission will attempt to land an "octacopter" on Saturn's moon Titan starting in 2027. These missions will all attempt to answer the question of whether we are alone. Mass spectrometry (MS) is a principal technique that scientists will rely on in spacecraft-based searches for ET life. MS has the advantage that it can simultaneously measure multitudes of compounds present in samples and thus provide a sort of "fingerprint" of the composition of the sample. Nevertheless, interpreting those fingerprints may be tricky. As best as scientists can tell, all life on Earth is based on the same highly coordinated molecular principles, which gives scientists confidence that all Earth life is derived from a common ancient terrestrial ancestor. However, in simulations of the primitive processes that scientists believe may have contributed to life's origins on Earth, many similar but slightly different versions of the particular molecules' terrestrial life uses are often detected. Furthermore, naturally occurring chemical processes are also able to produce many of the building blocks of biological molecules. Since we still have no known sample of alien life, this leaves scientists with a conceptual paradox: did Earth life make some arbitrary choices early in evolution, which got locked in, and thus life could be constructed otherwise, or should we expect that all life everywhere is constrained to be exactly the same way it is on Earth? How can we know that the detection of a particular molecule type is indicative of whether it was or was not produced by ET life? It has long troubled the scientists that biases in how we think life *should* be detectable, which are largely based on how Earth life is at present, might cause our detection methods to fail. *Viking 2* in fact returned odd results from Mars in 1976. Some of the tests it conducted gave signals considered positive for life, but the MS measurements provided no evidence for life as we know it. More recent MS data from NASA's *Mars Curiosity* rover suggest there *are* organic compounds on Mars, but they still do not provide evidence for life. A related problem has plagued scientists attempting to detect the earliest evidence for life on Earth: how can we tell if signals detected in ancient terrestrial samples are from the original living organisms preserved in those samples or derived from contamination from the organisms that presently pervade our planet? Scientists at the Earth-Life Science Institute at the Tokyo Institute of Technology in Japan and the National High Magnetic Field Laboratory (The National MagLab) in the United States decided to address this problem using a combined experimental and machine learning computational approach. The National MagLab is supported by the US National Science Foundation through NSF/ DMR-1644779 and the State of Florida to provide cutting-edge technologies for research.

Further Reading

Ghose, D., Das, N.K., Sinha, B. et al. (1994). Heat and helium release from thermal springs and influence of volcanic eruption. *Proceedings of the Indian National Science Academy* 60 (2): 481–486.

Guttenberg, N., Chen, H., Mochizuki, T., and James Cleaves, H. (2021). Classification of the biogenicity of complex organic mixtures for the detection of extraterrestrial life. *Life*. https://doi.org/10.3390/life11030234.

Hirose, T., Kawagucci, S., and Suzuki, K. (2011). Mechanoradical H_2 generation during simulated faulting: implications for an earthquake-driven subsurface biosphere. *Geophysical Research Letters* 38: L17303.

Hurowitz, J.A., Tosca, N.J., McLennan, S.M., and Schoonen, M.A.A. (2007). Production of hydrogen peroxide in Martian and Lunar soils. *Earth and Planetary Science Letters* 255: 41–52.

Love, J.J., Lucas, G.M., Murphy, B.S., et al. (2021). Down to Earth with nuclear electromagnetic pulse: realistic surface impedance affects mapping of the E3 geoelectric hazard. *Earth and Space Sciences* https://doi.org/10.1029/2021EA001792.

Mukherjee, S. (2001). Seismogenic potentiality of Delhi using remote sensing, soil geochemistry and geophysical data. *Indian Geological Congress* 2 (1) *(Spl.Vol.on Seismicity)*: 289–297. Research Highlight in Earth System science, DST, GOI. dst.gov.in/scientific-programme/ser-pac-es.htm.

Mukherjee, S. (2006). Influence of Starflare on the Sun-Earth environment and its possible relationship with snowfall. *European Geological and Geophysical Sciences Letter, (Germany) EGU* (14): 14–17. ISSN 1027-6343, www.the-eggs.org.

Mukherjee, S. (2020). Effect of Star-Burst on Sun-Earth environment vision and voyages for planetary science in the decade 2013–2022. *The National Academic Press USA*. https://solarsystem.nasa.gov/studies/41/effect-of-star-burst-on-sun-earth-environment/ https://www.nap.edu/catalog/13117/vision-and-voyages-for-planetary-science-in-the-decade-2013-2022 https://doi.org/10.17226/13117 ISBNs:Paperback: 978-0-309-22464-2 E book: 978-0-309-20957-1.

Mukherjee, S. (2021). Pixel based change to mitigate pandemics. *Journal of Earth Science and Technology* 2 (1): 1–2. ISSN: 2719-4094 http://www.htpub.org/Journal-Of-Earth-Sciences-And-Technology/.

Mukherjee, S. and Radovanović, M. (2011). IJErS11101). Influence of the Sun in the genesis of tornadoes. *The IUP Journal of Earth Sciences* January: 1–16.

Mukherjee, S. and Singh, P. (2014). Investigation of tectonic processes in the lunar South Polar Region using Mini-SAR and other data. *Frontiers in Earth Science* 2: 6. https://doi.org/10.3389/feart.2014.00006.

Mukherjee, S. and Singh, P. (2015). Identification of tectonic deformations on the South Polar surface of the Moon. *Planetary and Space Science* 112 (2015): 46–52.

Mukherjee, S., Radovanović, M., Stevančević, M. et al. (2011). Astrophysical analysis of Earthquake near Kraljevo (Serbia) on 03 November 2010. *Journal of the Geographical Institute "Jovan Cvijić" SASA* 61 (3): 1–15.

Singh, P. and Mukherjee, S. (2020). Chemical signature detection of groundwater and geothermal waters for evidence of crustal deformation along fault zones. *Journal of Hydrology* 582 (2020).

Index

Page locators in **bold** indicate tables. Page locators in *italics* indicate figures.

Extraterrestrial Remote Sensing and Climate Change, First Edition. Saumitra Mukherjee.

d

e

t

Printed and bound by CPI Group (UK) Ltd, Croydon, CR0 4YY
21/09/2026
14965814-0003